# Approximation Methods for High Dimensional Simulation Results – Parameter Sensitivity Analysis and Propagation of Variations for Process Chains

Inaugural - Dissertation

zur

Erlangung des Doktorgrades

der Mathematisch-Naturwissenschaftlichen Fakultät

der Universität zu Köln

vorgelegt von

Daniela Steffes-lai

aus Solingen

2014

Berichterstatter/-in:   Prof. Dr. Ulrich Trottenberg
(Gutachter)

                 Prof. Dr. Caren Tischendorf

Tag der mündlichen Prüfung: 15.01.2014

Bibliografische Information der Deutschen Nationalbibliothek

Die Deutsche Nationalbibliothek verzeichnet diese Publikation in der
Deutschen Nationalbibliografie; detaillierte bibliografische Daten sind
im Internet über http://dnb.d-nb.de abrufbar.

ISBN 978-3-8325-3696-1

Logos Verlag Berlin GmbH
Comeniushof, Gubener Str. 47,
10243 Berlin
Tel.: +49 (0)30 42 85 10 90
Fax: +49 (0)30 42 85 10 92
INTERNET: http://www.logos-verlag.de

# Zusammenfassung

In dieser Arbeit wird die Analyse mehrerer, aufeinanderfolgender Prozesse adressiert. Diese Prozessketten sind insbesondere bei der Herstellung einzelner robuster Produktkomponenten von großer Bedeutung. Unsicherheiten der Designparameter in jedem einzelnen der Prozessschritte können Auswirkungen auf die nachfolgenden Schritte haben, da dadurch die Anfangsbedingungen für folgende Schritte geändert und insbesondere Verteilungen von resultierenden Größen, wie Spannungen, inhomogen werden können.

Aufgrund des hohen Zeitaufwandes für hochaufgelöste Simulationen und eines enormen Datenaufkommens ist die ganzheitliche Betrachtung von Unsicherheiten in Prozessketten, d.h. insbesondere die Übertragung relevanter Schwankungen von einem Prozessschritt auf den nächsten, bisher nicht Stand der Technik.

Daher wird in dieser Arbeit die Methodik PRO-CHAIN entwickelt, welche eine effiziente Analyse, Quantifizierung und Übertragung der Einflüsse von Parameterunsicherheiten für komplette Prozessketten erlaubt.

Eine der wichtigsten Komponenten der neuen Methodik ist das entwickelte Parameterklassifizierungsverfahren. Dieses unterteilt Designparameter nach ihrer relativen Wichtigkeit und der Art des Einflusses auf das Ergebnis. Zusätzlich können mit der entwickelten Klassifizierung lokale Effekte des Prozesses identifiziert werden. Darauf basierend kann die Vorhersagegenauigkeit in diesen lokal interessanten Gebieten verbessert werden.

Durch ein iteratives Erstellen einer geeigneten Datenbasis von Simulationsergebnissen und eine physikalische Kompression dieser Datenbasis wird der Aufwand für folgende Analyseschritte minimiert, der Speicherverbrauch reduziert, und gleichzeitig eine geeignete Genauigkeit bei nichtlinearem Parametereinfluss sichergestellt.

Die komprimierte Datenbasis wird für die Vorhersage einzelner Zielfunktionale genutzt. Dazu werden Metamodelle mit radialen Basisfunktionen durch die komprimierte Datenbasis beschleunigt, so dass Vorhersagen lokal auf dem gesamten Bauteil effizient ermöglicht werden. Zudem wird eine Erweiterung dieser Modelle zur erfolgreichen Vorhersage der Versagensinitiierung für Crash Prozesse entwickelt.

Auf Basis der eingeführten Vorhersagemodelle wird eine lokale Approximation der Wahrscheinlichkeitsverteilung ermöglicht. Dadurch können, zum Beispiel, der Median und weitere Quantile als Robustheitsmaße in einer Optimierung lokal auf dem gesamten Bauteil betrachtet werden.

Zudem wird ein Schätzer des erwarteten Approximationsfehlers in einem vorhergesagten Zielfunktional in einem einzelnen Prozessschritt entwickelt, der direkt innerhalb der neuen Methodik berechnet wird, so dass dieser als aussagekräftiger Fehlerschätzer in den Anwendungen betrachtet werden kann. Ergänzend wird ein Schätzer für den maximalen Approximationsfehler für die Verkettung mehrerer Prozessschritte theoretisch hergeleitet.

Insgesamt ermöglicht die Methodik PRO-CHAIN die Übertragung nicht nur einzelner Ergebnisse, sondern aller relevanten Einflüsse von Parameterunsicherheiten von einem Prozessschritt auf den nächsten. Es wird gezeigt, dass dadurch die Vorhersagequalität des letzten Prozessschrittes deutlich verbessert werden kann.

Die neu entwickelte Methodik PRO-CHAIN wird auf industrielle Fragestellungen angewandt. Unter anderem wird eine komplexe Prozesskette von der Blechumformung hin zur Unfallsimulation (Crash) in der Automobilentwicklung untersucht. Damit werden die Effizienz sowie die Vorteile der Methodik im Vergleich zu typischen quasi-Monte Carlo Verfahren dargestellt. Insbesondere wird demonstriert, wie wichtige Einflüsse der Unsicherheiten identifiziert und lokale Effekte charakterisiert werden können, und ein mögliches Bauteilversagen korrekt vorhergesagt werden kann.

# Abstract

This work addresses the analysis of a sequential chain of processing steps, which is particularly important for the manufacture of robust product components. In each processing step, the material properties may have changed and distributions of related characteristics, for example, strains, may become inhomogeneous. For this reason, the history of the process including design-parameter uncertainties becomes relevant for subsequent processing steps.

In view of a still high simulation runtime of many physical processes and a huge amount of data to be analyzed, a comprehensive consideration of uncertainties in process chains is not state-of-the-art. Especially, the propagation of all relevant influences due to design-parameter variations from one processing step to the next has not been efficiently possible so far.

To address this, we have developed a methodology, called PRO-CHAIN, which enables an efficient analysis, quantification, and propagation of uncertainties for complex process chains locally on the entire mesh.

A parameter classification procedure newly developed is one of the essential components of the methodology proposed. It characterizes the design-parameters involved with regard to their relative importance and the nature of their influences on the design criteria. Additionally, local effects of the process can be identified using the classification approach. This enables the improvement of the forecast quality in these local regions of interest.

The new methodology introduces an iterative procedure for the extension of the database of simulation results in the case of nonlinear design-parameters to ensure a suitable accuracy. In combination with a compression of the database, this minimizes the computational costs and the memory requirements of subsequent analysis steps.

The compressed database is used in order to predict the behavior of new designs. For this purpose, metamodels with radial basis functions are accelerated by the ensemble compression, which makes the local prediction of the distributions aimed at efficiently possible. Furthermore, we have generalized these forecast models in order to enable the prediction of failure initiation in crash processes.

In conclusion, a local approximation of the probability distribution function becomes possible. Therewith, for example, the median and additional quantiles can be taken into account on the entire component in an optimization task in order to achieve a robust design.

Moreover, we have derived an estimator of the average approximation error in a predicted new design in a single processing step, which is computed directly within the new methodology. An additional estimator of the maximal approximation error for the entire process chain has been derived theoretically.

Bringing all individual components together, the newly developed methodology enables the propagation not only of single results, but also of all relevant influences of design-parameter uncertainties from one processing step to the next. We demonstrate that a considerably better forecasting quality of the final processing step is achieved using the new methodology.

The PRO-CHAIN methodology developed is applied to industrial applications from the automotive industry. Particularly, a complex forming-to-crash process chain is investigated. This demonstrates the efficiency and the benefits of the methodology proposed compared with state-of-the-art quasi-Monte Carlo methods. Especially, we illustrate that important influences due to design-parameter variations have been detected, local effects have been characterized, and a possible failure of the component has been predicted properly using the innovative methodology.

# Contents

# Chapter 1

## Introduction

### 1.1  Context

In recent years, product components have become more complex, while the product development periods have become ever shorter, especially in the automotive industry. In product development, the best possible product design has to be found with respect to multiple, often competing, design objectives, e.g., total weight, production costs, safety issues.

These design objectives are dependent on a range of design-parameters relating to defining features such as material properties and process configuration. Each design-parameter may be subject to variations. These variations can be process driven or caused by measurement errors. For example, suppliers ensure that the thickness of a metal sheet is within tolerances specified in a DIN norm, but it can vary within this range; friction coefficients are not fully controllable; temperature may depend on external influences like an open door on the factory floor. Such variations of design-parameters may have a substantial influence on the production process and the resulting products.

Figure 1.1 shows a clear example of strong influences of design-parameter variations on the final component behavior. In this example, extreme simulation results of a component crash test are shown, which arise due to variations in design-parameters [25, 112]. In particular, a strong interplay between a crack – strongly visible on the left-hand side – and a dent – strongly visible on the right-hand side – is observed.

Design-parameter optimization aims at achieving a robust design, which means that small variations in the design-parameters will only have a minimal impact on the results. Therefore, all effects of these variations have to be analyzed and quantified in order to achieve a robust design.

Figure 1.1: Extreme simulation results caused by design-parameter variations show a strong interplay between a crack and a dent.

Furthermore, product components are often manufactured by a sequential chain of processing or analysis steps. An example from the automotive industry is the forming- or casting-to-crash process chain. In each processing step, the material properties may have changed and distributions of related characteristics, for example, strains, may become inhomogeneous. For this reason, the history of the process becomes relevant for subsequent processing steps. For example, the forming of a metal sheet results in non-uniform distributions of thickness and strains and a preliminary damage of the sheet may occur.

Taking these inhomogeneous distributions and corresponding variations into account in a subsequent processing step is still not state-of-the-art. Specifically, the forming history including the local preliminary damage distribution is usually not taken into account in a crash analysis. That is, the last processing step is commonly still considered separately. While it is becoming standard practice to include at least first information from the process history, variations are omitted.

However, we have to consider all these inhomogeneous distributions in the subsequent processing step in order to get a realistic model. Therefore, a transfer of the results of the previous processing step to the subsequent processing step is essential. In addition, design-parameter variations lead to variations in the result, as already mentioned. For this reason, also a propagation of variations from one processing step to the next becomes necessary. Not only the results of the nominal simulation run, but also the entire range of results due to design-parameter variations has to be considered in order to allow a robust design to be obtained. Hence, it is necessary to analyze entire process chains and, furthermore, to get a complete insight into the local behavior of the process. The information gained will substantially improve the forecasting quality of single processing steps and, hence, of the ultimate product.

Simulation is used to reflect the physical effects of a real-world process over time in a virtual environment. The history of forming and crash simulations performed in the automotive industry dates back to the 1970s. In this time, simple, single component crashes were first investigated [109] and the application of the finite element (FE) method made the approximation of sheet metal processes possible [97, 119]. The first full vehicle crash simulation [109] became possible in 1986 predominantly by the upcoming of supercomputers. From that time on, the impact of crash simulations on automotive product development has changed rapidly. Crash simulations have started to be used in the product development and principal decision making process since the early nineties [120, 109] for reasons of the high performance gain due to the enabling of parallel computation, substantial improvements of simulation codes, and the possibility to reduce the time for process design. In the meantime, simulation has become an integral part of today's product development. Finely resolved meshes are necessary to model physical effects precisely. In addition, the meshes are usually refined adaptively, wherever the error of the current solution is high.

Since the end of the nineties, in combination with the rapid increase in computer power, it has become obvious that the reality given by uncertainties in the design-parameters, varying manufacturing conditions, and influences of the process, has to be considered. Especially, it has been experienced that changes of a certain design-parameter may influence the result substantially [109]. In the last years, analysis methods of sensitivity and robustness have been used increasingly. However, there are still discrepancies between the virtual and the real process behavior [52, 97]. Therefore, improvements in material models and the treatment of design-parameter uncertainties are particularly necessary in order to achieve a better simulation quality.

Many challenges have to be met when analyzing entire process chains. First, simulations may have meshes with thousands or millions of nodes, and hundreds of timesteps are performed. Several vector-valued simulation results, in combination with their variations distributed on the mesh, are of interest in order to achieve a robust design. This results in a huge amount of high dimensional data, which has to be analyzed. All analysis methods of high dimensional data will suffer from the curse of dimensionality [8], that is, their computational cost grows exponentially as a function of the dimension. Second, the simulation runtime of many physical processes is still high, while the development periods are decreasing. In addition, problems due to memory restrictions may occur, since each simulation run may provide several gigabytes of data. In view of these conditions, the efficient analysis of entire process chains in industrial applications has been difficult, if not impossible so far.

In conclusion, there is a high industrial need for efficient analysis methods of entire process chains including design-parameter uncertainties. See, for example, the future requirements for a virtual product development in the automotive industry specified in [50, 52, 97]. In particular, there have not existed any efficient methods to propagate all relevant influences due to design-parameter variations from one mesh to the mesh of the subsequent processing step.

To address this, we have developed a methodology for the efficient numerical analysis of process chains, called PRO-CHAIN. The methodology proposed enables the investigation of important dependencies between design-parameters, their variations, and design objectives locally on the entire mesh. Moreover, it allows the achievement of a robust design of single processing steps and entire process chains by a subsequent robust optimization. For this purpose, the methodology includes a fast and accurate prediction of the behavior of new designs, incorporating distributions of, for example, thicknesses, strains and damages, by means of a metamodel. Therewith, the statistical analysis of process chains, involving design-parameter uncertainties becomes possible.

Using the methodology developed, the problems due to the curse of dimensionality are partly avoided by reducing the dimension of the design-parameter space and minimizing the required number of simulation runs. Altogether, the PRO-CHAIN methodology leads to an intensive reduction of required memory and computational time compared with state-of-the-art quasi-Monte Carlo methods. For this reason, the PRO-CHAIN methodology becomes practicable in complex industrial applications.

## 1.2   Main Focus and Structure

The newly developed **PRO-CHAIN methodology** is an efficient approach for the numerical analysis of entire process chains that allows a robust design to be achieved. The methodology consists of different analysis steps addressing design-parameter sensitivity, dimension reduction, and robustness objectives. PRO-CHAIN can be applied to a wide range of applications arising from a variety of industrial fields. Exemplary applications in automotive engineering include the metal forming- or casting-to-crash process chains, and chains of several forming steps. Applications from the semiconductor industry include the process-to-device, and the device-to-circuit simulation chain.

Each component of the methodology can be used for single processing steps as well as process chains consisting of two and more sequential processes. The methods are based on properly constructed ensembles of simulation results stored in a database. However, the methodology developed is not restricted to numerical simulations. It can also be applied to analyze the behavior of real-world experiments, and to improve simulation models describing these experiments.

A general overview of the new PRO CHAIN methodology exemplified by the metal forming-to-crash process chain is illustrated in Figure 1.2 and described in the remaining of this section. The first processing step is a metal forming process, which describes the forming of a metal sheet into a component of a vehicle. The following, second processing step is a crash process of the formed component.

The analysis of the process chain starts with the investigation of the first processing step, in this example, a metal forming process. The design-parameters involved, in combination with their realistic variation ranges, have to be specified. For this reason, physical experiments with specimens and components for obtaining information about design-parameters used in material models and their realistic variations have to be performed and analyzed. In this thesis, we suppose that this has already been done.

Based on this information, a suitable **design of experiments (DoE)** is generated to determine an **ensemble of simulation runs (Chapter 3 and Chapter 4)**. The DoE is created so that the number of simulation runs is minimized. Then, the ensemble of forming simulation runs is carried out.

The result of each simulation run consists of several criteria, defined in each mesh node or mesh element. We consider the criteria sheet thickness, effective plastic strain (EPS), and preliminary damage in the following. All simulation results are stored in a **database** from which information about variations and distributions of the criteria on the entire mesh of the forming step can be generated.

Thereafter, the essential parts of the analysis of each processing step are performed. These parts of the PRO-CHAIN methodology deal with solving the problems of analyzing high dimensional data, that is, the curse of dimensionality, and restrictions in memory and computational time. Therefore, we have developed a design-parameter classification approach and a fast and accurate method to predict the behavior of new designs. The forecast method is based on a metamodel using radial basis functions (RBF's). This enables the investigation of different scenarios avoiding additional time-spending simulation runs. The combination of these methods make the efficient analysis of entire process chains practicable, also for complex non-linear industrial applications.

To be more specific, the **design-parameter classification approach** developed assigns each design-parameter to several importance classes, cf.

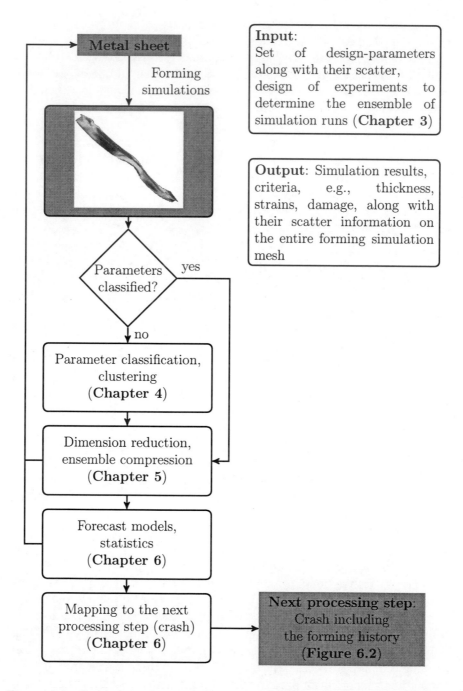

Figure 1.2: General overview of the PRO-CHAIN methodology exemplified by the metal forming-to-crash process chain.

**Chapter 4**. This procedure is based on a **sensitivity analysis**, which determines the influences of design-parameter variations on the criteria. On the one hand, the approach focuses on a design-parameter space reduction. On the other hand, it identifies design-parameters having a strongly nonlinear influence on the criteria. The influences of these design-parameters should be investigated in detail. Thus, additional simulation runs might be required in order to enable the generation of high-quality forecast models. State-of-the-art design-parameter sensitivity methods usually provide only global sensitivity measures. Moreover, their high computational costs make them applicable solely in combination with fast metamodels. The new approach proposed supports the fully local computation of sensitivity measures so that important **local regions of interest** are identified, which is of high importance for locally acting processes, like the forming process. This results in a good **estimate of the expected average prediction quality** of corresponding forecast models. Additionally, a **clustering** is derived based on the nonlinearity measure computed. This enables the local improvement of the forecast quality by applying more sophisticated approximation methods in the nonlinear domains.

Then, the results of the design-parameter classification are used to process the database so that the application of a forecasting method of the criteria becomes accurate enough, which is described in **Chapter 5**. This processing of the database includes a **dimension reduction** of the design-parameter space, an **iterative extension** of the database due to nonlinear effects, and an **ensemble compression** of the entire database. All information about variations relevant to the next processing step can be computed from the compressed database.

Thereafter, the behavior of new designs, which have not been simulated so far, can be predicted within the PRO-CHAIN methodology by means of metamodeling with RBFs. We have **accelerated** state-of-the-art **RBF metamodels** by an ensemble compression of the constructed (and propagated) database. For this ensemble compression, a singular value decomposition (SVD) of the database is used, see **Chapter 5**. This enables a fast **prediction of complete components** consisting of their geometry and resulting distributions of the criteria considered. Additionally, the further development of the metamodels focus on the **local prediction of failure** occurring in crash processes, as described in **Chapter 6**, which is essential to assess the results of such processes. For this purpose, an approach to deal with deleted mesh elements has been developed. Such mesh elements occur in crash simulations when a fracture is initiated, which leads to a crack in the component. Moreover, the fast forecast models enable the **computation of statistical information**, for example, quantiles of the criteria, on the entire mesh. This is necessary for a subsequent robustness analysis of processes with nonlinear local effects.

The next essential step of the PRO-CHAIN methodology is the **mapping procedure** outlined in **Chapter 6**. A strategy to choose the relevant information to be mapped is introduced that employs the forecast models derived. The number of simulation runs required is minimized using this strategy. Then, state-of-the-art mapping methods are used. In the mapping procedure, all relevant scatter information from one step is propagated to the mesh of the next step. In the example considered, the distributions of thickness, effective plastic strain, and preliminary damage along with their scatter information are mapped locally to the mesh of the crash step. This mapping becomes necessary because the mesh of the crash step is usually much coarser than the mesh of the forming step, and, in addition, some parts might be cut off. The mapping enables the usage of the **history of the process** within the current processing step. For example, different values of sheet thickness corresponding to the forming simulation result, are given in each mesh node instead of a global value for all nodes. This **improves the simulation forecasting quality** substantially, as we show in **Chapter 7**.

In the example considered, the second processing step is a crash process. The analysis of this subsequent step starts with the performance of an ensemble of crash simulation runs due to an appropriate DoE. As already stated, the history of the process serves as input for the crash simulation runs. Specifically, the distributed criteria of the forming step, that is, distributed thicknesses, effective plastic strains, and preliminary damages along with their scatter information are provided. The essential PRO-CHAIN methodology components, namely the design-parameter classification and ensemble compression of the database, are applied to this crash step, too. On the one hand, applying these parts of the PRO-CHAIN methodology is necessary in order to prepare the database to be used as input of a possible subsequent processing step. On the other hand, these parts are essential to construct a forecast model of the criteria of this step in order to enable the computation of statistics necessary to allow a robust design to be obtained. Since the history of the process is already considered in the crash step, the design-parameter classification and the forecast model are now based on the forming and crash step together. Thus, the entire process chain from the forming to the crash process is investigated.

The prediction of the behavior of new designs can be applied in each processing step. In the forming analysis, this enables the prediction of the technical feasibility of a forming process with the chosen design-parameter set, especially, in an early design stage. In the crash step, the crash behavior is predicted so that possible failures in a crash, like cracks and dents, can be detected. Moreover, statistical information on the criteria, like quantiles of the probability distribution, can be predicted with the PRO-CHAIN methodology, and, thus, used as robustness measures.

Furthermore, we derive an estimator of the maximal prediction error in a single processing step as well as for the entire process chain theoretically in order to **control the approximation quality**, in **Chapter 6**. An estimator of the average prediction error is additionally derived avoiding further computational effort to measure the maximal error in an arbitrary mesh node. Hence, it gives a valuable error estimate useful within numerical applications.

In summary, all these parts of the PRO-CHAIN methodology provide a major contribution to the **reduction of computational time** and **memory requirements** when analyzing single processing steps. Altogether, this enables the consideration and robust design of entire process chains in industrial applications.

To our knowledge, all relevant information about variations in order to approximate the probability distribution function have not been investigated and propagated between complex and finely resolved simulation meshes, so far. Therefore, as far we know, this is the first time that entire process chains including design-parameter uncertainties are analyzed. Particularly, the consideration of the complete history of a process and its variations within the simulation becomes possible. This can be used, together with the approximation of statistical information, for a robust optimization of the last processing step.

The remainder of this thesis is structured as follows.

**Chapter 2** introduces definitions and fundamentals which are important for the overall understanding of the methodology developed. In particular, the formal components of a process chain are defined.

**Chapter 3** reviews state-of-the-art methods with regard to different aspects of the methodology developed. It focuses on the applicability of standard methods in the context of high dimensional data and points out their limitations.

**Chapter 4** presents in detail the design-parameter classification procedure developed that aims at reducing the design-parameter space and at locating local regions of interest. The advantages in comparison with state-of-the-art sensitivity analysis methods are discussed.

**Chapter 5** derives the iterative construction of the database containing an ensemble of simulation results which reflects local variations of the output quantities considered. It focuses on the minimization of the computational effort for subsequent analysis steps.

**Chapter 6** introduces the acceleration of RBF metamodels, which allows their usage as forecast models for the criteria considered on the entire mesh. It derives a generalization of these metamodels enabling the prediction of failure in crash processes. Moreover, it presents a fast way to compute statistics of the metamodel response. Then, it introduces a strat-

egy for the appropriate propagation of all relevant scatter information to
the next processing step. Finally, the prediction errors are investigated.
Particularly, a prediction error formula for a single processing step as well
as for the entire process chain is derived theoretically.

**Chapter 7** presents benchmarks and industrial applications, which demonstrate the benefits and the efficiency of the innovative methodology
developed. A comparison of individual PRO-CHAIN components with
state-of-the-art methods illustrates that the new methodology is far more
efficient. A complex forming-to-crash process chain, by courtesy of Daimler AG, demonstrates that the new methodology enables the propagation of
relevant scatter from one processing step to the next resulting in a considerably improved forecasting quality of the crash result.

Finally, **Chapter 8** concludes this work and gives an outlook on further
research directions.

**Remark 1.1.** *Parts of this work have been published in conference contributions and papers [24, 47, 23, 22, 25, 111, 112, 12, 113, 115]. The
methodology newly developed is subject of the patent [114].*

# Chapter 2

---

# Notation and Fundamentals

---

Robust and optimal design-parameter settings under realistic conditions should be achieved in the product development process, even if many involved design-parameters are subject to variations. For this reason, all design-parameter variations have to be taken into account and transfered over the several processing steps to design the component considered in order to achieve a satisfactory result.

A design is called a **robust design** if small changes in the design-parameters, i.e., the initial conditions, will only have minimal impact on the results. To investigate robustness, we will focus on the distribution around the median of the solution, instead of the tails of the probability density function. The tails represent the mostly improbable cases, or worst case scenarios. Such scenarios are investigated in a **reliability analysis**, which is beyond our scope.

The influences of design-parameter variations on the solution are investigated in a **sensitivity analysis**. For this purpose, realistic ranges of variations have to be specified. As opposed to sensitivity analysis, **stability analysis** investigates very small variations, which can be, for example, numerical differences due to parallel computing (permutation of compute nodes) or round-off errors (different computer architectures). We focus primarily on sensitivity analysis assuming the influences due to variations of design-parameters are much higher than the ones due to numerical variations. This assumption is correct in the examples considered.

In this chapter, we introduce some fundamental terms and definitions which are important for the overall understanding of the process chains concept. In particular, we introduce the notation of the formal components of a process chain in **Section 2.1**. Then, in **Section 2.2**, some fundamentals and general approaches, including statistics, interpolation, and mapping of

11

data, are briefly repeated. In general, the reader is assumed to be familiar with these concepts. In **Section 2.3**, a brief introduction in forming processes is given, since one focus of the numerical examples throughout this thesis is on forming processes.

## 2.1   Terminology

### Design-Parameters and Criteria

We investigate the influence of inputs (design-parameters) and their variations (scatter) on outputs (criteria) which are mainly high dimensional simulation results in this thesis. Both, design-parameters and criteria, are defined in each node of (different) discrete meshes. Design-parameters are, e.g., material parameters (anisotropy, density, damage coefficients), geometry, process parameters (friction, forces), or external influences (open door in the factory floor). Criteria are usually functionals on the entire mesh, for example, thicknesses, strains, stresses, and (preliminary) damages, or scalar optimization criteria, like total mass, maximal force, and energy.

We denote the number of design-parameters by Npar and the **design-parameter space** by $\mathbb{R}^{\mathrm{Npar}}$. A member of the design-parameter space is a **design-parameter** vector $\mathbf{P} = [P^1, \ldots, P^{\mathrm{Npar}}]^T$. For brevity, we denote design-parameters with *parameters* in the following. Analogously, we define a **criteria space** $\mathbb{R}^{\mathrm{Ncrit}}$ with dimension Ncrit and denote a member of this space by **criteria** vector $\mathbf{Y} = [Y^1, \ldots, Y^{\mathrm{Ncrit}}]^T$. A criterion can either be an analytical function value, a physical measurement, or a simulation result. In this thesis, we mainly focus on simulation results defined in each mesh node ($N_i$) or mesh element ($\mathrm{Element}_i$). The number of mesh nodes is defined as Nnodes. The number of **timesteps** of a simulation run is denoted by Nts. A particular timestep is called **state** and denoted by ts.

Formally, a certain **processing step** in an arbitrary mesh node $N_i$ can be described by a function $g_i \colon \mathbb{R}^{\mathrm{dim}+\mathrm{Npar}} \to \mathbb{R}^{\mathrm{dim}+\mathrm{Ncrit}}$, where dim is the dimension of the coordinate space, usually $\mathrm{dim} = 3$. Since we do not change the given discretization, i.e., the input coordinates $\mathbf{X}^I$, we define $g_i(\mathbf{P}) := g_i(\mathbf{X}^I, \mathbf{P})$, with $\mathbf{X}^I \in \mathbb{R}^{\mathrm{dim}}$ and $\mathbf{P} \in \mathbb{R}^{\mathrm{Npar}}$. The corresponding solution is defined as $\widetilde{\mathbf{Y}} := (\mathbf{X}, \mathbf{Y}) = g_i(\mathbf{P})$ with $\mathbf{X} \in \mathbb{R}^{\mathrm{dim}}$ and $\mathbf{Y} \in \mathbb{R}^{\mathrm{Ncrit}}$.

The overall processing step is described on the entire mesh as the function vector $\mathbf{g} := [g_1, \ldots, g_{\mathrm{Nnodes}}]^T$ with the mesh nodes $N_1, \ldots, N_{\mathrm{Nnodes}}$. Thus, the parameter vector $\mathbf{P}$ has to be specified in each mesh node. Considering the first processing step of a process chain, $\mathbf{P}$ is usually the same in each mesh node. Then, the evaluation of each component of $\mathbf{g}$ leads to criterion vectors $\mathbf{Y}_1, \ldots, \mathbf{Y}_{\mathrm{Nnodes}}$, with $\mathbf{Y}_i = \left[Y^1, \ldots, Y^{\mathrm{Ncrit}}\right]^T$. The output coordinates $\mathbf{X}$ are usually the input coordinates $\mathbf{X}^I$ with displacements.

**Remark 2.1.** *The relationship between parameters and criteria is often described by a system of partial differential equations (PDEs) or a set of functions. Due to the fact that this relationship is usually not provided analytically in the applications considered, but a black-box function, we restrict the formal description to one single function* $\mathbf{g}$.

**Example 2.2.** *We write* $\mathbf{g}_1$ *for the first and* $\mathbf{g}_2$ *for the second processing step in a common process chain consisting of two steps. Hence, we consider* $\mathbf{g}_2 \circ \mathbf{g}_1(\mathbf{P})$. *In Section 7.3, we consider the industrial application of a forming (U) to crash (C) process chain* $\mathbf{g}_C \circ \mathbf{h} \circ \mathbf{g}_U(\mathbf{P})$, *where* $\mathbf{h}$ *represents an auxiliary function for dimension reduction and mapping tasks.*

In practice, $\mathbf{g}$ is usually a black box, and only a set of discrete function values in several sampling points in each mesh node is given. A finite set of Nexp distinct **sampling points** in the parameter space is represented by $\mathrm{S} := \{\mathbf{P}_1, \ldots, \mathbf{P}_{\mathrm{Nexp}}\}$, defined in each mesh node. The corresponding function values (criteria) are given by $\{\mathbf{Y}_1, \ldots, \mathbf{Y}_{\mathrm{Nexp}}\}$ in each mesh node, too. In general, the cardinality of a finite set $\Xi$ equals the number of elements of the set and is denoted by $\#\Xi$, e.g., $\#\mathrm{S} = \mathrm{Nexp}$. In the following, for brevity, we refer to the set of sampling points and corresponding function values known as *the data given*. Points which do not belong to the set of sampling points are referred to as **out-of-sample** points. The evaluation of an out-of-sample point, for example, by means of a metamodel, is the prediction of the behavior of a **new design**.

**Remark 2.3.** *In this thesis, we deal with simulation results reflecting a specific processing step. Therefore,* $\mathbf{g}$ *is a black box and the discrete function values given are the simulation results in the mesh nodes, unless stated otherwise. For example, in a two dimensional criteria space with the two criteria, simulation result effective plastic strain (EPS) and simulation result sheet thickness (t), the criteria vector* $\mathbf{Y} = [Y^1, Y^2]^T = [EPS, t]^T$ *is given in each mesh node.*

We construct two types of **databases**, one which contains only information about a single criterion and a single simulation timestep for all sampling points, and another one which contains all criteria and, additionally, coordinates and optionally several timesteps for all sampling points. The first database is a matrix $\mathbf{M} \in \mathbb{R}^{\mathrm{Nnodes} \times \mathrm{Nexp}}$ with entries $M_{kj}$, which consists of a fixed criterion $Y^i \in \{Y^1, \ldots, Y^{\mathrm{Ncrit}}\}$ for all samplings in each node, i.e.,

$$\mathbf{M} := \begin{pmatrix} Y^i_{1,1} & Y^i_{1,2} & \cdots & Y^i_{1,\mathrm{Nexp}} \\ Y^i_{2,1} & Y^i_{2,2} & \cdots & Y^i_{2,\mathrm{Nexp}} \\ \vdots & \vdots & \ddots & \vdots \\ Y^i_{\mathrm{Nnodes},1} & Y^i_{\mathrm{Nnodes},2} & \cdots & Y^i_{\mathrm{Nnodes},\mathrm{Nexp}} \end{pmatrix}. \tag{2.1}$$

The latter database contains the information on the states block by block, in which each block contains the coordinate and criteria information $\widetilde{\mathbf{Y}}$ for all nodes, that is, $\widetilde{\mathbf{M}} \in \mathbb{R}^{\text{Ndata} \times \text{Nexp}}$, with $\text{Ndata} = \text{Nts}(\text{Nnodes}(\dim + \text{Ncrit}))$ and

$$\widetilde{\mathbf{M}} := \begin{pmatrix} \widetilde{\mathbf{M}}(1) \\ \widetilde{\mathbf{M}}(2) \\ \vdots \\ \widetilde{\mathbf{M}}(\text{Nts}) \end{pmatrix}. \tag{2.2}$$

$\widetilde{\mathbf{M}}(i) \in \mathbb{R}^{(\text{Nnodes}(\dim + \text{Ncrit})) \times \text{Nexp}}$ belongs to the $i$-th state and is defined as

$$\widetilde{\mathbf{M}}(i) := \begin{pmatrix} \mathbf{X}_{1,1} & \mathbf{X}_{1,2} & \cdots & \mathbf{X}_{1,\text{Nexp}} \\ \mathbf{Y}_{1,1} & \mathbf{Y}_{1,2} & \cdots & \mathbf{Y}_{1,\text{Nexp}} \\ \mathbf{X}_{2,1} & \mathbf{X}_{2,2} & \cdots & \mathbf{X}_{2,\text{Nexp}} \\ \mathbf{Y}_{2,1} & \mathbf{Y}_{2,2} & \cdots & \mathbf{Y}_{2,\text{Nexp}} \\ \vdots & \vdots & \ddots & \vdots \\ \mathbf{X}_{\text{Nnodes},1} & \mathbf{X}_{\text{Nnodes},2} & \cdots & \mathbf{X}_{\text{Nnodes},\text{Nexp}} \\ \mathbf{Y}_{\text{Nnodes},1} & \mathbf{Y}_{\text{Nnodes},2} & \cdots & \mathbf{Y}_{\text{Nnodes},\text{Nexp}} \end{pmatrix}. \tag{2.3}$$

In most industrially relevant applications, the parameters are not known exactly, but can vary considerably. The variation of these parameters can have crucial influences on the production process and, thus, on the quality of the resulting product. Therefore, we have to formulate the parameters $P^i$ as stochastic variables, so that $\mathbf{P}$ becomes a random vector. We denote the **nominal parameter vector** with $\mathbf{P}_{\text{nom}} := [P^1_{\text{nom}}, \ldots, P^{\text{Npar}}_{\text{nom}}]^T$, where $P^i_{\text{nom}}$ can be the mean value of the parameter $P^i$ or an arbitrary, but fixed value of $P^i$ within the parameter space. We refer to the simulation run with nominal parameter set as *nominal simulation run*. We define parameter **variation/scatter** in two ways. For each parameter a variation/scatter might be given as standard deviation from its nominal value, denoted by $\sigma_{P^i}$. This means that the parameter can take each value in the interval $[P^i_{\text{nom}} - \sigma_{P^i}, P^i_{\text{nom}} + \sigma_{P^i}]$. In practice, we often do not have information on the probability distribution of a parameter. Then, we define lower and upper bounds of the parameter. In this case, the parameter can take each value in $[P^i_{\text{min}}, P^i_{\text{max}}]$. These bounds might come from DIN standards. With $P^i_{\text{nom}} = (P^i_{\text{min}} + P^i_{\text{max}})/2$, we can choose $\sigma_{P^i} = (P^i_{\text{max}} - P^i_{\text{min}})/2$. The vector of standard deviations of all parameters is defined by $\sigma_{\mathbf{P}} := [\sigma_{P^1}, \ldots, \sigma_{P^{\text{Npar}}}]^T$. The minimum parameter vector is defined as $\mathbf{P}_{\text{min}} := [P^1_{\text{min}}, \ldots, P^{\text{Npar}}_{\text{min}}]^T$ and the maximum parameter vector is given by $\mathbf{P}_{\text{max}} := [P^1_{\text{max}}, \ldots, P^{\text{Npar}}_{\text{max}}]^T$.

**Remark 2.4.** *If we consider scatter, $\mathbf{P}$ will become a random vector. In this case, the criteria $Y^i$ are stochastic variables and the criteria vector $\mathbf{Y}$ will become a random vector as well. Therefore, the databases, given by*

*Equations (2.1) and (2.2), each represent an* **ensemble** *of solutions, i.e., realizations of the random criteria. This ensemble can be used for further analysis, e.g., the computation of statistical information.*

We will use the following notation throughout the thesis:

$$\mathbf{P}^{(i)} := [P^1, \ldots, P^{i-1}, P^{i+1}, \ldots, P^{\text{Npar}}]^T, \tag{2.4}$$

$$g(\mathbf{P}^{i\pm1}) := g(P^1, \ldots, P^{i-1}, P^i \pm \sigma_{P^i}, P^{i+1}, \ldots, P^{\text{Npar}}), \tag{2.5}$$

$$g(\mathbf{P}^{i\pm1,j\pm1}) := g(P^1, \ldots, P^{i-1}, P^i \pm \sigma_{P^i}, P^{i+1}, \ldots,$$
$$P^{j-1}, P^j \pm \sigma_{P^j}, P^{j+1}, \ldots, P^{\text{Npar}}). \tag{2.6}$$

## Meshes

In this thesis, the finite set of mesh nodes $\mathbf{X}_i \in \mathbb{R}^{\text{dim}}$ usually originates from a finite element (FE) discretization resulting in unstructured, irregular meshes. A mesh is represented by **nodes**, **elements**, and a list of **connectivities**. It represents a certain geometry, e.g., a component part of a car.

The number of mesh nodes in a domain $\Omega$ is denoted by $\text{Nnodes}_\Omega$. The elements used are mainly quadrilaterals and triangles. $\text{Con}(\text{Element}_i)$ denotes the **connectivity list** of the $i$-th element, which lists all nodes belonging to this element.

In many simulation codes, the mesh is refined adaptively during the simulation procedure. This means that an element is subdivided into several smaller elements, wherever an error indicator is high. An **adaptive mesh refinement** is used to improve the accuracy of the simulation result. For this reason, mesh based results of different simulation runs cannot be compared directly. A transfer of results to a common reference mesh becomes necessary. This procedure is explained briefly in Section 2.2.3.

## Norms

Let the $L_p$-norms for a vector $\mathbf{X} \in \mathbb{R}^{\text{dim}}$ be denoted by

$$||\mathbf{X}||_p := \left( \sum_{i=1}^{\text{dim}} |X_i|^p \right)^{1/p}, \quad 1 \le p < \infty. \tag{2.7}$$

For $p = 2$ we get the Euclidean norm. For $p \to \infty$ the $L_p$-norm approaches the maximum norm denoted by

$$||\mathbf{X}||_\infty := \max_{i=1,\ldots,\text{dim}} |X_i|. \tag{2.8}$$

In the case of real matrices $\mathbf{A} \in \mathbb{R}^{m \times n}$, we use a matrix norm which is consistent with the given vector norm. In particular for square matrices $\mathbf{A} \in \mathbb{R}^{n \times n}$ , we make use of the *induced matrix norm* of the Euclidean norm ($p = 2$) which is given by the spectral norm $||\mathbf{A}||_2$ and defined as

$$||\mathbf{A}||_2 := \sqrt{\alpha_{\max}(\mathbf{A}^T \mathbf{A})}, \qquad (2.9)$$

where $\alpha_{\max}$ denotes the maximal eigenvalue. If $\mathbf{A}$ is symmetric, then

$$||\mathbf{A}||_2 = \max_i \left\{ |\alpha_i| \ : \ \alpha_i \text{ eigenvalue of } \mathbf{A} \right\} =: |\alpha|_{\max} \qquad (2.10)$$

holds. Moreover, we consider the Frobenius norm $||\mathbf{A}||_F$ given by

$$||\mathbf{A}||_F := \sqrt{\sum_{i=1}^{m} \sum_{j=1}^{n} a_{ij}^2} = \sqrt{\text{trace}(\mathbf{A}^T \mathbf{A})}. \qquad (2.11)$$

## 2.2    Fundamentals and General Approaches

In this section, some fundamentals and general approaches are briefly repeated. This includes fundamentals of probability theory and statistics (**Section 2.2.1**), interpolation (**Section 2.2.2**), and the transfer of data, called mapping, between distinct meshes (**Section 2.2.3**). Some mapping approaches for specific scenarios are explained additionally in **Section 2.2.3**, because this data preprocessing is fundamental to understanding the PRO-CHAIN methodology. Especially, determining a reference mesh, and mapping between mesh element based and mesh node based values is described.

### 2.2.1    Stochastics

We repeat some definitions and statements of stochastics (see a textbook on stochastics, like [96]) that we use in the following chapters. Let $(\Sigma, \mathcal{A}, \mathcal{P})$ be a probability space, with event space $\Sigma$, $\sigma$-algebra $\mathcal{A}$, and probability measure $\mathcal{P}$.

The **probability distribution function** of a real-valued random variable $Z$ is defined as

$$F_Z(s) = \mathcal{P}[Z \leq s]. \qquad (2.12)$$

The corresponding **probability density function** for continuous random variables is defined as

$$f_Z(s) = F_Z'(s). \qquad (2.13)$$

The **expectation** of a random variable $Z$ indicates the *expected, average* or *mean* value of $Z$ and is defined as

$$\mathbf{E}[Z] = \int_{-\infty}^{\infty} z f_z dz. \tag{2.14}$$

If we have a sampling $\{Z_1, \ldots, Z_{\mathrm{Nexp}}\}$ from $Z$, the expectation can be approximated by its *sample mean*

$$\mathbf{E}[Z] = \frac{1}{\mathrm{Nexp}} \sum_{i=1}^{\mathrm{Nexp}} Z_i. \tag{2.15}$$

The **variance** of $Z$ is given by

$$\mathbf{Var}[Z] = \mathbf{E}[(Z - \mathbf{E}[Z])^2] = \mathbf{E}[Z^2] - \mathbf{E}[Z]^2, \tag{2.16}$$

and $\sqrt{\mathbf{Var}[Z]}$ is called the **standard deviation**. The **covariance** of two random variables $W, Z$ is given by

$$\mathbf{Cov}[W, Z] = \mathbf{E}[(W - \mathbf{E}[W])(Z - \mathbf{E}[Z])] = \mathbf{E}[WZ] - \mathbf{E}[W]\mathbf{E}[Z]. \tag{2.17}$$

A $q$-**quantile** for a random variable is defined as the value $z$, such that the probability that the random variable will be less than $z$ is at most $q$, with $q \in [0, 1]$, that is,

$$\mathbf{Q}_q(Z) = F_Z^{-1}(q) = \inf\{z | \mathcal{P}[Z \leq z] \geq q\}. \tag{2.18}$$

Let $Z_{(1)} \leq \cdots \leq Z_{(\mathrm{Nexp})}$ be the order statistics of a sampling $\{Z_1, \ldots, Z_{\mathrm{Nexp}}\}$, so that the samples are sorted in increasing order. A $q$-quantile is generally estimated from this order statistics, see, e.g., [27]. A standard estimator of the $q$-quantile is, for example, given by

$$\widehat{\mathbf{Q}}_q(Z) = Z_{(\lceil q\mathrm{Nexp}\rceil)}, \tag{2.19}$$

where $\lceil z \rceil$ is the integer ceiling of $z$. In order to improve the precision of the quantile estimator, methods based on the weighted sum of the order statistics have been developed, for example, [45, 103]. The $0.5$-quantile is called **median** and separates the higher half of a distribution from the lower half.

A frequently used distribution function is the **normal distribution** $N(\mathbf{E}[Z], \sigma^2)$ (*Gaussian distribution*) with variance $\mathbf{Var}[Z] = \sigma^2$ and density function

$$f_Z(z) = \frac{1}{\sqrt{2\pi\sigma^2}} \exp\left(-\frac{(z - \mathbf{E}[Z])^2}{2\sigma^2}\right). \tag{2.20}$$

A normal distribution with zero expectation ($\mathbf{E}[Z] = 0$) and unit variance ($\sigma^2 = 1$) is called *standard normal distribution* and is denoted by $\Phi_Z$ with density function $\varphi_Z$.

## 2.2.2   Interpolation

**Metamodels**, also called *response surface models*, are widely used to approximate the dependencies between parameters and criteria, based on a set of sampling points.   Let an unknown or computationally expensive scalar function $g_i$ be determined by a set of Nexp distinct sampling points $S := \{\mathbf{P}_1, \ldots, \mathbf{P}_{\text{Nexp}}\} \subset \mathbb{R}^{\text{Npar}}$ in the parameter space with corresponding function values $\{g_i(\mathbf{P}_1), \ldots, g_i(\mathbf{P}_{\text{Nexp}})\}$. Then, an approximation $\widetilde{g}_i$ for $g_i$ is sought, which is faster to evaluate, so that the approximation condition

$$\sum_{j=1}^{\text{Nexp}} \|g_i(\mathbf{P}_j) - \widetilde{g}_i(\mathbf{P}_j)\|_2 \to \min, \tag{2.21}$$

holds. If we additionally require

$$\widetilde{g}_i(\mathbf{P}_j) = g_i(\mathbf{P}_j), \quad \forall j = 1, \ldots, \text{Nexp}, \tag{2.22}$$

then, the discrete sampling points are interpolated. Without loss of generality, we consider scalar-valued functions $g_i \colon \mathbb{R}^{\text{Npar}} \to \mathbb{R}$. Vector-valued functions $\mathbf{g} \colon \mathbb{R}^{\text{Npar}} \to \mathbb{R}^{\text{Ncrit}}, \text{Ncrit} > 1$ can be interpolated component-by-component.

There are many commonly used approximation and interpolation methods, like linear regression models, polynomial interpolation, metamodels, kriging, and artificial neural networks. Metamodels are widely used, for example, in the field of optimization (for an overview see [84, 32, 65, 60]), and are discussed in more detail in Section 3.4. We use metamodels with radial basis functions (RBFs) as forecast model in order to predict the behavior of new designs, that is, the results of out-of-sample points, within the PRO-CHAIN methodology. We have developed these RBF metamodels further in order to deal with failures, which is important within the applications considered, see Section 6.1.

Additionally, there exist interpolation methods specializing in solving stochastic partial differential equations, for example, the stochastic Galerkin and collocation methods, see Section 3.5. A comparison between the new methodology and a stochastic collocation method is given in Section 7.1 for a model problem.

## 2.2.3   Mapping

Let $\Omega_I \subset \mathbb{R}^{\text{dim}}$ be the **input/source domain** with mesh nodes $\mathbf{X}_i^I \in \Omega_I, i = 1, \ldots, \text{Nnodes}_I$. Analogously, let $\Omega_O \subset \mathbb{R}^{\text{dim}}$ be the **output/target domain** with mesh nodes $\mathbf{X}_i^O \in \Omega_O, i = 1, \ldots, \text{Nnodes}_O$. Note that these two meshes might have a different resolution or/and geometry. For example, the number of nodes might be different due to adaptive refinement of the mesh within the simulation or parts of the source mesh are cut off in the target mesh. The transfer of data, more precisely an arbitrary quantity $v_i^I = v(\mathbf{X}_i^I)$ on the source mesh to a quantity $v_i^O = v(\mathbf{X}_i^O)$ on the target mesh, is called **mapping**. We can think of $v^I, v^O$ as functions $v^I, v^O : \mathbb{R}^{\dim} \to \mathbb{R}$, then the **database of functionals** on the meshes is given by

$$\mathbf{v^I} = (v_1^I, \ldots, v_{\mathrm{Nnodes}_I}^I)^T \in \mathbb{R}^{\mathrm{Nnodes}_I \times \dim}, \qquad (2.23)$$

$$\mathbf{v^O} = (v_1^O, \ldots, v_{\mathrm{Nnodes}_O}^O)^T \in \mathbb{R}^{\mathrm{Nnodes}_O \times \dim}. \qquad (2.24)$$

Mapping between non-matching meshes becomes necessary in several scenarios. First, mapping enables using data provided at one processing step as input data of the next processing step, which performs on a different mesh, e.g., due to different discretization, in the case of several processing steps connected in a process chain. For instance, the crash meshes are usually much coarser as the forming meshes in a forming-to-crash process chain. Second, a fixed reference mesh has to be specified in a single processing step with multiple simulations with adaptive mesh refinement. Then, all data of each simulation should be mapped to this reference mesh in order to create a comparative database with a single discretization.

Since the simulations considered in the applications are finite element codes, we suppose finite dimensional function spaces $V^I, V^O$ as subspaces of appropriate function spaces, e.g., $L^2(\Omega_I), L^2(\Omega_O)$, and corresponding basis functions $\psi_i^I, \psi_i^O$, so that (cf. [54, 57])

$$v^I = \sum_{i=1}^{\mathrm{Nnodes}_I} v_i^I \psi_i^I, \quad \{\psi_i^I\} \subset V^I \subset L^2(\Omega_I), \qquad (2.25)$$

$$v^O = \sum_{i=1}^{\mathrm{Nnodes}_O} v_i^O \psi_i^O, \quad \{\psi_i^O\} \subset V^O \subset L^2(\Omega_O). \qquad (2.26)$$

The mapping problem is defined to determine the $v_i^O$ given the $v_i^I, V^I$ and $V^O$. There exist many mapping algorithms in the literature, mainly based on interpolation and extrapolation. Accurate approximations can be achieved by minimization of the mapping residual $v^O - v^I$ in an appropriate norm (e.g., the $L_2$-norm of a comparison space) and by as much conservation of physical properties as possible, measured by the integral of the residual. A detailed discussion of mapping techniques is beyond our scope.

**Remark 2.5.** *For mapping between different processing steps, we use the software MpCCI MetalMapper®* [35] *by Fraunhofer SCAI, which is based on a modified Shepard algorithm [90].*

---

*MpCCI MetalMapper® is a registered trademark of Fraunhofer Institute for Algorithms and Scientific Computing SCAI.

To distinguish between several mapping scenarios, we define the following **mapping functions**:

$$m_j : \text{Mapping within a single processing step } j. \qquad (2.27)$$
$$m_{ij} : \text{Mapping between distinct processing steps } i \text{ and } j. \qquad (2.28)$$

In Equation (2.27) the target mesh is a reference mesh for all simulation runs of this processing step. In the mapping in Equation (2.28) the data is transfered from the resulting mesh of one processing step $i$ to the input mesh of the subsequent processing step $j$.

### Reference Mesh

The meshes of the different simulation runs according to parameter variations remain the same as long as the mesh is not adaptively refined during the simulation.

Often, specifically in metal forming simulations, adaptive mesh refinement is used to improve the accuracy of the simulation results. Therefore, starting with the same mesh, simulation runs with different parameter sets result in adaptively refined, and thus, different meshes. This implies that the number of resulting mesh nodes differs and that simulation results are not defined on the same mesh nodes anymore. Thus, results of different simulation runs cannot be compared directly. In this case, a fixed **reference mesh** has to be specified. The results of each simulation corresponding to the criterion vector are transferred to this reference mesh. After mapping, all extracted functionals have the same dimension and can be compared node per node.

There are several possibilities to specify a reference mesh. For example, the mesh of the last timestep of the nominal simulation run including its local refinements is a suitable choice. We prefer to choose the nodes of an earlier state as reference. In this state, several adaptive refinement steps should be performed already, so that areas of high interest have an appropriate resolution. Nevertheless, the meshes should not be highly different yet. For example, the state immediately before a fracture is initiated can be chosen. The exact timestep from which a reference mesh is chosen depends on the application. The mapping can be performed by means of any appropriate approximation method, as stated above.

### Mapping Between Element Based and Node Based Values

All analysis and prediction methods of the PRO-CHAIN methodology are based on data which can be represented as point clouds, that is, nodes given with coordinates and assigned nodal values. Therefore, all criteria stored in the databases, given by Equations (2.1) or (2.2), have to be assigned to nodes. However, some criteria, for example, the damage information, are computed element based in the forming and crash simulations. For this reason, a mapping between element based and node based values becomes necessary. Criterion values assigned to nodes are referred to as **nodal values**, and the values assigned to elements are referred to as **element values**, respectively.

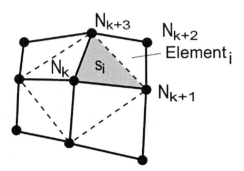

Figure 2.1: Scheme to compute the value of node $N_k$ from the contributing element values. The surface area $s_i$ of the triangle computed from Element$_i$ contributing to $N_k$ is colored in gray.

The algorithm used for mapping between element and nodal values is described in the following. A general scheme to compute the value of node $N_k$ from the contributing element values is depicted in Figure 2.1. The approximated criterion value $n_k$ assigned to a node $N_k$ is the sum of contributing element values $e_i$ weighted with the surface areas $s_i$ of corresponding triangles. For example, when computing the value of $N_k$, a contributing value $e_i$ of Element$_i$ is weighted by the surface area of the triangle spanned by the vectors $\overrightarrow{N_{k+1} - N_k}$ and $\overrightarrow{N_{k+3} - N_k}$, as shown in Figure 2.1.

In general, the value of a node $n_k$ is computed from the contributing element values by

$$n_k = \frac{\sum_{i, N_k \in \text{Con}(\text{Element}_i)} e_i s_i}{\sum_{i, N_k \in \text{Con}(\text{Element}_i)} s_i}, \tag{2.29}$$

where Con(Element$_i$) denotes the connectivity list of the $i$-th element. Since the nodes of an element are numbered counter clockwise, the surface area $s_i$ assigned to Element$_i$ contributing to $N_k$ is given by the Euclidean norm of the cross product of the assigned vectors, that is,

$$s_i = \frac{1}{2} \|(\overrightarrow{N_{k+1} - N_k}) \times (\overrightarrow{N_{k+3} - N_k})\|_2. \tag{2.30}$$

Each simulation result assigned to elements is mapped with the algorithm described to nodal values in order to be used within the PRO-

CHAIN methodology. Specifically, the databases, given by Equations (2.1) and (2.2), already store nodal based criteria computed with this mapping.

## 2.3 Sheet Metal Forming Processes with Example

**Sheet metal forming** is a production process, where the intended shape of a component is created through plastic deformation of an initial workpiece, usually a sheet metal blank. Furthermore, no material is removed in this process.

We demonstrate several parts of the PRO-CHAIN methodology by a fundamental forming example. Furthermore, the industrial application considered in Section 7.3 includes a metal forming process as first processing step. Therefore, we give a short overview of sheet metal forming processes to gain a general understanding of the parameters involved, the resulting criteria, and the dependencies between them.

**Plastic deformation** describes permanent shape change, whereas elastic deformation is reversible, so that the original shape is returned, if the load is removed. A general introduction to metal forming and a detailed overview of its mechanics is given in the textbooks [122, 79, 5]. There are

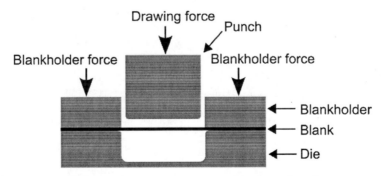

Figure 2.2: Schematic representation of a deep drawing process.

a lot of common sheet metal forming techniques. We concentrate on **deep drawing processes**, which are frequently used in the automotive industry, and to which the applications considered belong to. Deep drawing is commonly performed with tools consisting of a punch, a die, and a blankholder as shown in Figure 2.2. A blankholder is used to prevent wrinkles. The blank has to slide between the die and the blankholder, thus the contact is usually realized by a lubricant. Therefore, the friction coefficient turns out to be an important process parameter strongly influencing the forming results.

**Formability** refers to the ability of a material to be shaped through plastic deformation without defects. Usual defects in deep drawing processes are wrinkling, orange skin, and fractures. The defects are either caused by the forming tools, the friction, the material properties, or geometrical parameters, cf. [79]. Formability can be evaluated by several methods based on mechanical as well as simulating tests, for details refer to [79, 5].

Since the 1970s there has been a rapid development of FE methods for sheet metal **forming simulations**. For an overview, state-of-the-art, and necessary future developments in sheet metal forming simulations refer to [97, 119]. Today, sheet metal forming simulations are widely used in the automotive industry for the verification of the technical feasibility in early design stages as well as for the optimization of the forming process. Still, further developments in order to improve the forming simulations are needed, especially, the material models including anisotropy and hardening laws need to be extended, e.g., for high-strength steels [97].

In the following, we briefly summarize some **material properties** in order to get a basic understanding of the parameters influencing the forming simulation results, for details refer to the textbooks [79, 5].

The total deformation is expressed in terms of **strains**. A deformation component can be written as the sum of an elastic and a plastic part, i.e.,

$$\varepsilon_{ij}^{\text{total}} = \varepsilon_{ij}^{\text{e}} + \varepsilon_{ij}^{\text{p}}, \tag{2.31}$$

where $\varepsilon^{\text{e}}$ denotes the *elastic strain* and $\varepsilon^{\text{p}}$ the *plastic strain*, respectively. In general, $\varepsilon$ denotes the *true strain* [79], which is defined as

$$\varepsilon = \int_{l_0}^{l} d\varepsilon = \int_{l_0}^{l} \frac{dl}{l} = \ln \frac{l}{l_0}, \tag{2.32}$$

where $l$ denotes the gauge length of the specimen in a tensile test. The *effective strain* denotes the total effect of all true strain, which is used to compare two different states of strain. The effective strain can be written as [122]

$$\bar{\varepsilon} = \sqrt{\frac{2}{3}\varepsilon_{ij}\varepsilon_{ij}} = \sqrt{\frac{2}{3}\left(\varepsilon_x^2 + \varepsilon_y^2 + \varepsilon_z^2 + 2\left(\varepsilon_{xy}^2 + \varepsilon_{yz}^2 + \varepsilon_{zx}^2\right)\right)}. \tag{2.33}$$

Analogously, the **effective plastic strain (EPS)** is the plastic part of the effective strain. The EPS is a monotonically increasing scalar value and can be computed with [74]

$$\bar{\varepsilon}^p = \int_0^T d\bar{\varepsilon}^p, \tag{2.34}$$

with $T$ the time and $d\bar{\varepsilon}^p = \sqrt{\frac{2}{3}d\varepsilon_{ij}^p d\varepsilon_{ij}^p}$.

Most metallic materials consist of an aggregate of crystal grains [122]. Plastic deformation can occur through dislocation movements of particles. The crystallographic texture and the rolling process lead to an **anisotropy** of mechanical properties of sheet metals [5], which means that the properties are strongly dependent on the orientation (rolling, diagonal, and transverse direction). Anisotropy is described by the anisotropy coefficient, called *Lankford parameter*, which is measured in different directions. The anisotropy coefficient $R$ is defined as the ratio of width strain ($\ln w/w_0$) and thickness strain ($\ln t/t_0$). The thickness strain can be calculated from the width and length using the constant volume assumption

$$wtl = w_0 t_0 l_0, \tag{2.35}$$

where $w_0$ and $w$ are the initial and final width, and $l_0$ and $l$ are the initial and final gauge length of the specimen in a tensile test. Then, the anisotropy coefficient can be expressed as

$$R = \frac{\ln \frac{w}{w_0}}{\ln \frac{t}{t_0}} \underset{(2.35)}{=} \frac{\ln \frac{w}{w_0}}{\ln \frac{w_0 l_0}{wl}}, \tag{2.36}$$

for details see [79, 5]. The index of the anisotropy coefficient indicates the direction in which $R$ is measured, that is, the angle between the axis of the specimen and the rolling direction, e.g., $R_0$, $R_{45}$, and $R_{90}$ for tests in the rolling, diagonal, and transverse directions, respectively.

The **plastic behavior** of a material can be described by a yield criterion, a flow rule, and a hardening law. When a material reaches the *yield strength* it passes from elastic into plastic behavior. This point can be determined with the corresponding stress-strain curve of the material, which can be derived from uniaxial tests, cf. [5, 79]. A typical stress-strain curve is shown in Figure 2.3. In general, a *yield criterion* [5] describes the conditions under which plastic deformation occurs in case of a multiaxial stress state. The yield criterion is expressed as an implicit function, often called *yield surface* or *yield locus*. Many yield criteria have been developed specialized either for isotropic materials or for anisotropic materials. A detailed overview of the historical development of yield criteria can be found in [5].

**Remark 2.6.** *We use the yield criterion proposed by Barlat (1989) [5] in the industrial application in Section 7.3. The Barlat 1989 yield criterion takes anisotropy into account, which is modeled by the Lankford parameters. It is implemented in the FE code LS-DYNA®[†], version 971, under material model 36 [75].*

---

[†]LS-DYNA® is a registered trademark of Livermore Software Technology Corporation.

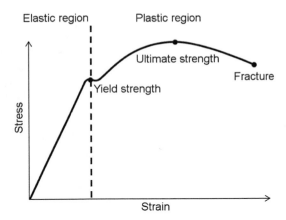

Figure 2.3: A typical stress-strain curve.

In the linear-elastic region stress is directly proportional to strain, that is, *Hooke's law* holds. In the plastic area the relationship between stresses and strains is described by a *flow rule*.

The plastic deformation causes a hardening, so that the material properties change. Therefore, a hardening rule has to be given, which describes the evolution of the yield surface during yielding. For example, the often applied *Swift hardening law* is given by

$$\bar{\sigma} = K(\varepsilon_0 + \bar{\varepsilon})^n, \tag{2.37}$$

where $\bar{\sigma}, \bar{\varepsilon}$ denote the effective stress and effective strain, respectively, $K$ denotes the strength coefficient, $n$ is the strain-hardening index, which defines the slope of the stress-strain curve at high strains, and $\varepsilon_0$ is a *pre-strain* constant, which indicates a shift in the strain axis if the material has been hardened already [79].

We introduce a fundamental example of sheet metal forming processes in the remaining of this section, which is used in the following chapters to demonstrate the developed concepts.

**Fundamental Example: Forming of a Pan with Secondary Design Elements**

A fundamental forming example is the deep drawing process of a pan geometry with secondary design elements. This model problem is used as benchmark problem in the following chapters in order to demonstrate several components of the PRO-CHAIN methodology developed. A pan geometry is a benchmark problem frequently used in the automotive and steel industry to investigate the general material behavior under forming. The model

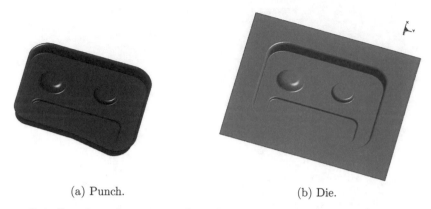

(a) Punch.                                          (b) Die.

Figure 2.4: Punch and corresponding die with secondary design elements.

problem considered in this work, provided by SIMUFORM GmbH, is one of
a set of pan geometries, which has been specially designed to investigate a
wide range of industrially relevant forming characteristics [116]. In particular, the influence of the secondary design elements to their surroundings
is considered. Furthermore, the model problem is designed to take different
kinds of possible curvatures into account, from flat areas to convex or con-
cave surfaces. We consider a deep drawing process of this model problem.
In the benchmarks performed, the simulation runs required are carried out
with the FE-code LS-DYNA®. The corresponding forming tools are shown
in Figure 2.4. Adaptive mesh refinement is used to get accurate simulation
results.

| Parameter type | $P^j$ | $\sigma_{P^j}$ |
|---|---|---|
| Material | Thickness $t$ [mm] | 2% |
| | Lankford parameter $R_{90}$ | 10% |
| | Strength coefficient in Swift law $K$ | 10% |
| | Strain hardening index in Swift law $n$ | 10% |
| Process | Friction coefficient $\mu$ | 50% |
| | Hold down force $F^H$ [N] | 10% |

Table 2.1: Involved parameters and their range of variation in pan forming
example.

We consider the variation of four material parameters and two process
parameters in the forming process. Table 2.1 shows the involved parameters
and their variation range. In particular, the variation of the parameters
of the Swift hardening law used to characterize the material behavior are
illustrated in Figure 2.5.

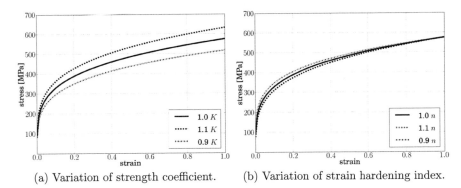

(a) Variation of strength coefficient.    (b) Variation of strain hardening index.

Figure 2.5: Influences of parameter variations on the Swift hardening law.

The resulting criteria considered in this example are the distribution of EPS and sheet thickness. The results of the nominal simulation run are shown in Figure 2.6.

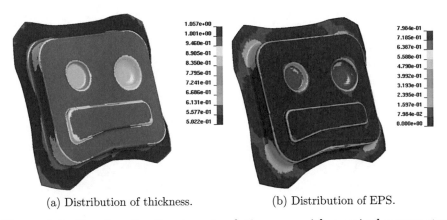

(a) Distribution of thickness.          (b) Distribution of EPS.

Figure 2.6: Results of a forming simulation run with nominal parameter set.

# Chapter 3

---

# Mathematical Concepts

---

Many different methods from different mathematical fields are needed in the context of analyzing high dimensional data efficiently. In this chapter, a survey on state-of-the-art methods with respect to different aspects of the PRO-CHAIN methodology is given in order to get a deeper understanding of the theoretical background. We focus on the advantages and drawbacks of each method. Moreover, we discuss the applicability of the methods with regard to analyzing high dimensional data. Some of the methods will be adopted within the PRO-CHAIN methodology, others seem not to be agreeable with the objectives of analyzing process chains effectively.

This chapter is organized as follows. **Section 3.1** deals with design of experiments (DoE) methods for determining an appropriate set of sampling points with respect to certain criteria. DoE methods are needed in several steps within the PRO-CHAIN methodology. For example, these methods are required to determine the points on which the data is collected, and the points in which a forecast model is evaluated in order to compute statistical information.

**Section 3.2** gives an overview of common approaches of sensitivity analysis, which aims at determining the parameters with the most influence on the response. In each processing step, many parameters are involved, which may be subject to variations. Using sensitivity analysis, we are able to specify the impact of each parameter on the criteria considered. Especially, parameters which have hardly any impact on the response can be neglected in further analysis. Hence, sensitivity analysis is an essential step to reduce the dimension of the parameter space in order to decrease the computational time of subsequent investigations.

**Section 3.3** reviews dimension reduction methods used to further reduce the model complexity. Together with a sensitivity analysis, this builds an important component to deal with the curses of dimensionality.

**Section 3.4** investigates approximation methods of the response, called metamodels, which are fast to evaluate. Especially, we focus on radial basis function (RBF) metamodels. These metamodels are used for the substitution of computationally expensive simulation runs.

**Section 3.5**, finally, introduces specialized approximation methods for solving stochastic partial differential equations. A numerical comparison between RBF metamodels and stochastic collocation methods is given in Section 7.1.


## 3.1   Design of Experiments

Design of experiment (DoE) methods address the choice of an appropriate set of sampling points $S := \{P_1, \ldots, P_{Nexp}\}$ with respect to certain objectives. At these sampling points the model $g$ is evaluated and corresponding criteria $\{Y_1, \ldots, Y_{Nexp}\}$ are generated, where model is a synonym for, e.g., an analytical function, a physical measurement, a simulation, or an approximation model. The result and the accuracy of data analysis highly depends on the data collected. This turns the DoE into an essential task in high dimensional data analysis. The optimal set of sampling points depends on the objective the DoE is used for.

In this thesis, we need DoE methods with regard to two main objectives. On the one hand, we have to determine an ensemble of simulation runs in order to construct a metamodel in each processing step (cf. Figure 1.2). On the other hand, DoE is used for further data analysis, e.g., to compute statistical information based on the metamodel. Especially, if DoE methods are used to construct a metamodel, space filling designs are required that allow a complete exploration of the design space. Even a uniform distribution of sampling points is often needed in order to get an appropriate metamodel accuracy. In each case, the corner points of the design space play an important role. If these points are not contained in the DoE, the evaluation of the metamodel in the corner regions of the design space will be most often also not allowed because this will result in extrapolation. Moreover, it should be possible to extend the generated DoE iteratively in order to improve the accuracy of metamodels by adding selected sampling points. In addition, the DoE method should be practicable for a large number of random parameters.

We discuss the main aspects of several commonly used DoE techniques under the above requirements, see, for example, [80, 98, 66]. A summary of DoE techniques particularly beneficial for metamodeling is given, e.g., in [65, 63, 104].

**Remark 3.1.** *The random parameters $P^j$ are usually called* factors *and the number of possible values each factor can take is called* levels *in DoE literature.*

In general, DoE methods are based on three main principles, namely *replication, randomization* and *blocking* in order to ensure randomly distributed output variables and to estimate the experimental error, see [80].

One of the first popular DoE methods is **Taguchi's parameter design approach**, which has been developed in the 1980s. This method strictly distinguishes between control factors and noise factors. Correspondingly, an inner *orthogonal array* consisting of the variations of the control factors and an outer orthogonal array for the noise factors is generated. These two arrays are crossed so that for each experiment from the inner array the noise factors are varied according to the outer array. Furthermore, Taguchi developed some measures of variation called *signal-to-noise ratios* (SNR), which are dependent on the design objective and take the response mean and variation into account.

The drawbacks of this method concern the separation of the control and noise factors, as well as the signal-to-noise ratios. These are discussed, for example, in [104, 84] and references therein. Especially, interactions between control factors cannot be taken into account by Taguchi's design procedure. Additionally, it may lead to a very large number of experiments required. Data with different mean and variance structure can lead to the same SNR, and are, thus, equivalent within the meaning of Taguchi's measures, which may lead to wrong conclusions.

The **one-factor-at-a-time experiment** is the most simple DoE approach, which selects a nominal parameter vector $\mathbf{P}_{\text{nom}}$. Then, for each parameter, only this parameter is varied within its range while the remaining parameters are kept constant [80]. The major limitation of this approach is that only linear dependencies can be explained, whereas interactions between parameters cannot be detected.

A special one-factor-at-a-time experiment is a **star-point DoE**, which consists of experiments for each parameter at three levels, the nominal parameter vector, the minimum, and the maximum parameter vector ($\mathbf{P}_{\text{min}}$, $\mathbf{P}_{\text{max}}$). In total $2\text{Npar} + 1$ experiments are performed.

**Remark 3.2.** *We return to the star-point approach in the context of sensitivity analysis in Section 3.2 and Chapter 4.*

A **full factorial design** is a basic DoE technique, which investigates all possible combinations of $k$ levels and Npar parameters resulting in $k^{\text{Npar}}$ sampling points.

Factorial designs allow the investigation of interactions among parameters. However, their major disadvantage is that the number of sampling points grows exponentially with the number of parameters. Therefore, factorial designs become impracticable in industrial applications, where Npar and $k$ are large.

To circumvent this drawback, **fractional factorial designs** have been developed, which perform only a subset of a factorial design. Refer to [80] for an overview of the class of factorial designs. These designs have in common to place sampling points at the corners of the design space so that the extreme scenarios are investigated.

A Latin square is a square grid containing sampling points, so that in each row and each column is one and only one point [80]. The **Latin hypercube sampling (LHS)** is a generalization of this concept to arbitrary dimensions. For that purpose, the range of each parameter is divided into Ndata bins of equal probability, so that several Latin squares are superimposed.

This concept does not guarantee a uniform distribution of the sampling points. Hence, optimal LHS designs with respect to the uniform distribution of the sampling points have been developed, cf. [65, 63].

LHS enables the generation of space filling designs with an arbitrary number of sampling points independent of the number of parameters. However, the design cannot be extended to additional sampling points. If it becomes necessary to add additional sampling points, we have to start from scratch with a complete new LHS design.

**Monte Carlo (MC)** methods generate random sampling points at which the model is evaluated. They reach a probabilistic error bound of $\mathcal{O}(1/\sqrt{\text{Nexp}})$. Therefore, despite the error bound is independent of the number of random variables Npar, a very high number of sampling points is required to get a low error. An extended overview of MC methods can be found, for example, in the textbook [33].

**Quasi-MC** methods [85, 86, 16] replace the random sampling points by deterministic, more uniformly distributed **low discrepancy sequences** of points. Discrepancy is a measure of the deviation from the uniform distribution. Examples of low discrepancy sequences are *Halton*, *Sobol* and *Faure* sequences, see, for example, [81, 16]. It has been shown that quasi-MC methods can reach a better performance and lower errors as MC methods for a wide range of applications, for example, numerical integration [81]. Under certain conditions, quasi-MC methods converge more rapidly with a rate of $\mathcal{O}(\text{Nexp}^{-1}(\log \text{Nexp})^{\text{Npar}})$ [16]. However, this convergence rate, and, thus, the advantage over MC methods is observable only for small dimensions and smooth functions [16, 91]. If Npar becomes large, the bound is dominated by the logarithmic term unless $\text{Nexp} > 2^{\text{Npar}}$.

**Remark 3.3.** *(Quasi-)MC is usually a means to obtain a reference solution in cases where no analytical solution is provided. In the following, we use* *quasi-MC with the Halton sequence as reference solution in statistical computations, unless stated otherwise. Moreover, (quasi-)MC methods are the standard approach for statistical analysis, e.g., the computation of quantiles, especially in engineering applications.*

**Sparse grids** are based on tensor product formulas of a one dimensional multiscale basis. For a detailed overview of sparse grids refer to [15] and references therein.

A common way to generate multidimensional sampling points is to use full tensor products of one-dimensional interpolating functions. Therewith, the number of required sampling points grows exponentially with the number of parameters. That is, a discretization with $k$ points in each direction leads to $Nexp = k^{Npar}$.

Sparse grids use instead the Smolyak construction [106], which is a linear combination of full tensor product formulas. This results in a construction with only $\mathcal{O}(k(\log k)^{Npar-1})$ sampling points, so that the curse of dimensionality is overcome to some extent. The current number of points selected depends on the Smolyak level $l$ used. The resulting sparse grid is given by the union of the pairwise disjoint grids of the one-dimensional formulas [37].

To generate the one-dimensional formulas different quadrature rules can be used. *Clenshaw-Curtis points*, which are the extreme points of the Chebyshev polynomials, or *Gauss points*, which are zeros of polynomials that are orthogonal with respect to the probability density function of the random variable, are commonly used.

Quadrature formulas are nested, if the lower-order rule uses a subset of the sampling points of the higher-order rule, i.e., $S_l^{Npar} \subset S_{l+1}^{Npar}$. This is an important property, which makes the iterative extension of samplings possible. A comparison of different quadrature formulas with respect to costs and accuracy is given in [37].

**Remark 3.4.** *Sparse grids are commonly used as DoE method in stochastic collocation methods. The construction of multidimensional collocation points by means of sparse grids can be found, for example, in [127, 89]. We discuss stochastic collocation methods in Section 3.5.*

## 3.2 Sensitivity Analysis

Parameters which are known only approximately can be modeled mathematically as stochastic variables. The resulting mathematical model of the problem includes random variables, that is, parameters which are subject to variations. For this reason, the response or output of the mathematical model is a random variable or random field as well (cf. Section 2.1, specifically, Remark 2.4).

Let $b$ denote a representative of a function $b \colon \mathbb{R}^{\mathrm{Npar}} \to \mathbb{R}^{\mathrm{Ncrit}}$.

**Definition 3.5** (Sensitivity analysis). *Let us consider a mathematical model $b$ having* Npar *independent random parameters* $\mathbf{P} = [P^1, \ldots, P^{\mathrm{Npar}}]^T$, *and a random criterion vector* $\mathbf{Y} = [Y^1, \ldots, Y^{\mathrm{Ncrit}}]^T$:

$$\mathbf{Y} = b(\mathbf{P}). \tag{3.1}$$

*A **sensitivity analysis** determines the change in the response of a model according to changes in parameters [48, 56, 99]. Hence, a sensitivity analysis aims to identify the parameters having the most influence on the criteria.*

The results of a sensitivity analysis can be used to reduce the parameter space to the most influencing parameters (cf. Chapter 4), and, thus, to reduce the model complexity. In the following, we consider Ncrit $= 1$ without loss of generality.

**Remark 3.6.** *A **stability analysis** investigates very small numerical variations. For example, numerical differences due to parallel computing caused by a permutation of compute nodes, and round-off errors due to different computer architectures are investigated. As already stated, we focus on sensitivity analysis. The influences due to variations of parameters are much higher than the ones due to numerical variations in the examples considered. Thus, we suppose the model investigated to be numerically stable.*

Sensitivity analysis methods can be divided into local methods, briefly described in Section 3.2.1, and global methods, summarized in Section 3.2.2. For a detailed discussion of local and global methods refer, for example, to [92, 99].

## 3.2.1   Local Methods

Local sensitivity methods [92, 121] are partial derivative based methods. This approach is motivated by the **Taylor series expansion** of the response $b$ around a nominal parameter set $\mathbf{P}_{\mathrm{nom}}$, which expresses the impact of a parameter variation on the response. A second-order Taylor series expansion of $b$ can be written compactly as

$$\begin{aligned} b(\mathbf{P}) =\,& b(\mathbf{P}_{\mathrm{nom}}) + (\mathbf{P} - \mathbf{P}_{\mathrm{nom}})^T \mathbf{J}(\mathbf{P}_{\mathrm{nom}}) \\ & + \frac{1}{2!}(\mathbf{P} - \mathbf{P}_{\mathrm{nom}})^T \mathbf{H}(\mathbf{P}_{\mathrm{nom}})(\mathbf{P} - \mathbf{P}_{\mathrm{nom}}) \\ & + \cdots \end{aligned} \tag{3.2}$$

with $\mathbf{J}$ the **Jacobian matrix** of $b$ and $\mathbf{H}$ the **Hessian matrix** of $b$. Note, with $\mathbf{P} = \mathbf{P}_{\max} = [P^1_{\max}, \ldots, P^{\mathrm{Npar}}_{\max}]^T$, the expression $\mathbf{P} - \mathbf{P}_{\mathrm{nom}}$ equals $\sigma_{\mathbf{P}}$ (cf. Section 2.1).

Thus, local methods analyze the influence of variations of arbitrary, but fixed parameter sets on the response instead of the uncertainty over the entire parameter space. The sensitivities computed around an arbitrary nominal parameter set cannot be carried over to other regions in the parameter space. Hence, local methods will be only reasonable, if a nominal parameter set is known, otherwise it will be a "guess" in the parameter space.

The partial derivatives can be approximated numerically by **finite differences (FD)**, which are derived by truncating the Taylor series expansion at a certain order. The local first order **sensitivity index** $J_j$ for parameter $P^j$ is defined as first order finite difference approximation of the Jacobian matrix. This approach is also known as a *"one-parameter-at-a-time"* experiment (cf. one-factor-at-a-time experiments, Section 3.1).

The choice of the step size $\sigma_{P^j}$ is essential, since first order sensitivity indices with FD approximations are only valid, as long as the first order Taylor series is valid around $\mathbf{P}_{\text{nom}}$. If $\sigma_{P^j}$ is chosen too high, the linearity assumption in a neighborhood of $\mathbf{P}_{\text{nom}}$ will fail. In this case, higher order terms in the Taylor series might have important influences on the response. A brief discussion of the choice of $\sigma_{P^j}$ is given in [92].

**Remark 3.7.** *If the parameters have different scales, the sensitivities will not be comparable directly. Therefore, normalized sensitivity indices $LS_j$ should be used. The normalization by the standard deviation of each parameter is an appropriate choice, leading to $LS_j = J_j \sigma_{P^j}$.*

Higher order indices can be derived with the same approach using finite differences of higher order derivatives. Only function evaluations at a set of Nexp distinct sampling points have to be performed, so that these methods are suitable for "black box" functions. Nexp = Npar + 1 evaluations are needed to obtain all first order sensitivity indices. Nevertheless, the computational effort to approximate higher order derivatives increases significantly, since more function evaluations are needed. Due to this reason, first order indices are usually computed only, cf. [121], which can only reflect the linear influences of the parameters on the response.

**Remark 3.8.** *In the context of reliability analysis, **first and second order reliability methods (FORM/SORM)** [93, 18] are usually applied to estimate the probability of failure. These methods are similar to local sensitivity methods based on first and second order partial derivatives. However, the methods are only applicable in the case of a linear or quadratic function dependence and normally distributed parameters.*

**Remark 3.9.** *Further local methods, like the **automatic differentiation** [43], **direct differential methods**, and their extension, the **Green's***

*function method [92, 121] exist. These methods require that the underlying function is known analytically. Thus, these methods are not pursued further in this thesis. This is due to the fact that mainly finite element simulation codes are used when analyzing process chains. Therefore, the underlying set of partial differential equations (PDEs), or the simulation source code, is not known or freely available. Instead, we have to treat the simulation as black box.*

### 3.2.2  Global Methods

Global sensitivity methods analyze the model by computing averaged sensitivity measures over the entire parameter space without specifying a fixed nominal parameter set. Thus, these methods take into account, a priori, the ranges or distribution functions of the parameters and explore the entire parameter space [92]. Global methods are especially suited for large variations of the parameters, where no nominal response of the problem is known. Thus, they aim for an understanding of the general parameter behavior on the response. A survey on global sensitivity methods is given, for example, in [99, 92, 121, 49].

Global sensitivity methods include response surface approximation or metamodeling techniques, described in Section 3.4. Especially, regression-based methods are widely used for sensitivity analysis [49].

Another commonly used approach is the averaging of **derivative-based measures** over the parameter space. That is, the global derivative based sensitivity index for parameter $P^j$ is given by $\mathbf{E}[J_j(\mathbf{P})]$. A well known derivative-based method is the **screening method by Morris** and extensions, which are discussed, e.g., in [70, 99]. Besides the mean, the standard deviation of the distribution function of the sensitivities is computed as indicator for interactions. A correlation between the Morris sensitivity indices and variance-based global sensitivity indices is established in [108].

Furthermore, **stochastic approaches** [26] and related methods, like the **Fourier Amplitude Sensitivity Test (FAST) method**, investigate the probability density function of the response. These methods usually have a high computational effort in order to construct the entire probability density function [92, 121].

Moreover, a main focus of global sensitivity methods is on **variance-based methods**. These methods decompose the variance of the response according to the variance of the parameters in order to determine which part of the variation in the response can be explained by the variation of which (group of) parameters [99, 107]. In the remaining of this section, we briefly review the state-of-the-art variance-based **Sobol sensitivity indices** [107], since these are most widely used in the literature.

Without loss of generality, let $I$ be the unit interval $[0, 1]$ and $P^i \in I, i = 1, \ldots, \text{Npar}$. Furthermore, let $b$ be a square integrable function. Then, the representation of $b$ in the form

$$b(\mathbf{P}) = b_0 + \sum_{s=1}^{\text{Npar}} \sum_{i_1 < \cdots < i_s}^{\text{Npar}} b_{i_1 \ldots i_s}(P^{i_1}, \ldots, P^{i_s}) \tag{3.3}$$

$$= b_0 + \sum_{i=1}^{\text{Npar}} b_i(P^i) + \sum_{i,j>i} b_{ij}(P^i, P^j)$$

$$+ \cdots + b_{12\ldots\text{Npar}}(P^1, \ldots, P^{\text{Npar}}),$$

where $1 \leq i_1 < \cdots < i_s \leq \text{Npar}$, $b_0$ a constant, is called *ANOVA-representation* (ANOVA - analysis of variance) of $b(\mathbf{P})$ according to Sobol [107], if

$$\int_0^1 b_{i_1 \ldots i_s}(P^{i_1}, \ldots, P^{i_s}) dP^k = 0 \quad \text{for} \quad k = i_1, \ldots, i_s. \tag{3.4}$$

Thus, the $2^{\text{Npar}} - 1$ non-constant summands in Equation (3.3) are orthogonal in pairs and can be expressed as integrals of $b$, details of the properties can be found in [107, 99]. By squaring Equation (3.3) and integrating over the parameter space, the total variance $D$ and the variances $D_{i_1,\ldots,i_s}, s = 1, \ldots, \text{Npar}$ can be derived

$$D_{i_1,\ldots,i_s} = \int b_{i_1,\ldots,i_s}^2(P^{i_1}, \ldots, P^{i_s}) \, dP^{i_1} \cdots dP^{i_s}, \tag{3.5}$$

$$D = \mathbf{Var}[b(\mathbf{P})] = \int b^2(\mathbf{P}) \, d\mathbf{P} - b_0^2$$

$$= \sum_{s=1}^{\text{Npar}} \sum_{i_1 < \cdots < i_s}^{\text{Npar}} D_{i_1 \cdots i_s}. \tag{3.6}$$

The **Sobol sensitivity indices** are defined as the ratios

$$S_{i_1,\ldots,i_s} = \frac{D_{i_1 \cdots i_s}}{D}, \tag{3.7}$$

where $s$ determines the order of the sensitivity index. These indices can be computed using the conditional expectations, see [99]. The first order indices $S_i$ measure the influence of each parameter $P^i$ taken alone. The higher order indices measure interaction effects of various parameters.

To measure the total contribution (first order plus all higher order interaction effects) of a parameter onto the response the **total sensitivity indices** $S_{T_i}$ have been developed [51, 117] as the sum over all partial sensitivity indices including parameter $P^i$.

Sobol's sensitivity indices can be computed numerically by MC based methods, for details and discussion refer to [107, 99, 51]. The computational effort of the MC based approach to compute all first order Sobol indices equals $\text{Nexp}(\text{Npar} + 2)$ simulation runs [99]. Due to a high computational effort, only the first order and total sensitivity indices are usually computed. Thus, a main drawback of variance-based methods is that they are highly computationally expensive. Due to that fact, this approach is hardly applicable to complex applications directly, such as finite element simulations.

Instead, the sensitivity indices can be computed based on metamodels of the response function, see, for example [117, 56]. For example, in [117] a polynomial chaos expansion of the response is determined. Based on this metamodel, the Sobol indices are derived analytically. Note that the derived sensitivity indices represent the sensitivities of the metamodel. For this reason, their accuracy depends on the metamodel accuracy.

**Remark 3.10.** *Building the metamodel, that is, performing the* $\text{Nexp}$ *simulations, is often already computationally expensive in the applications considered. In this case, a possible approach is to reduce the number of parameters by another, e.g., derivative-based approach. Afterwards, the variance-based sensitivity method can be applied to a reduced set of parameters. In this case, the Sobol sensitivity analysis will offer only an additional benefit, if the detailed parameter behavior, especially the higher order interactions, are of interest.*

## 3.3   Dimension Reduction Methods

Dimension reduction methods aim at embedding a high dimensional data set into a subspace of lower dimension. This lower dimension is determined to be the intrinsic dimension of the data set. The **intrinsic dimension** of a data set is defined as the minimal number of parameters required to represent the data set sufficiently [73, 17, 64]. Thus, reducing the data set to its intrinsic dimension can avoid the problems due to the curse of dimensionality to some extend.

Most dimension reduction methods require the intrinsic dimension as a fixed input parameter. However, the estimation of the intrinsic dimension is usually computationally expensive. Moreover, there are only few methods on this estimation in the literature, mostly related to the fractal dimension. These methods include the box counting or capacity dimension, the information dimension, and the correlation dimension [42, 64, 17]. In few dimension reduction methods, an estimator of the intrinsic dimension is already integrated, for example, in the principal component analysis (PCA) (described in Section 3.3.2), and its local variants [36, 61].

The transformation of the data set to an intrinsic-dimensional subspace is the purpose of dimension reduction methods. A possible approach to reduce the complexity of the data set is to find similarities among the data points. Then, only one representative of each group with respect to the similarities found is sufficient in order to describe the properties of the data set completely. Finding such disjoint subsets of the data set is the objective of clustering algorithms, which are briefly described in Section 3.3.1.

Thereafter, classical dimension reduction methods are investigated. Linear dimension reduction methods suppose that the data set results from a linear transformation of variables. The advantage of linear methods is that they are usually very simple and fast to compute. The PCA, as a representative of linear dimension reduction methods, is introduced in Section 3.3.2.

Nonlinear dimension reduction methods make no assumption on the construction of the data. Thus, these methods are more powerful to detect nonlinear relationships among the data as linear dimension reduction methods. A survey on and a discussion of nonlinear dimension reduction methods is given in Section 3.3.3.

## 3.3.1  Clustering

Clustering denotes the task to find disjoint subsets, called *clusters*, of data points within a given data set, so that points within a cluster are similar. Then, only the set of one representative of each cluster, for example, the cluster centers, has to be investigated in order to analyze the properties of the entire data set.

**Remark 3.11.** *Clustering is often denoted by* **vector quantization clustering** *in the field of compression, dimension reduction, and learning algorithms.*

We briefly summarize the **K-means algorithm** as a simple to implement and widely used clustering algorithm. We use it for the clustering of the data based on the results of the parameter classification procedure, described in Section 4.2.

### K-means Clustering

K-means [62, 77] denotes an iterative clustering algorithm to find $K =$ Nclust clusters $CL_1, \ldots, CL_{Nclust}$ for Ndata data points. Each cluster is represented by its *cluster center*, which is given by the mean vector $c_{CL_k}$, $k = 1, \ldots,$ Nclust over all points belonging to that cluster. K-means determines the clusters, so that the mean squared distance from each data point $y_i$ to its nearest cluster center is minimized [77], that is,

$$\sum_{k=1}^{\text{Nclust}} \sum_{y_i \in \text{CL}_k} ||y_i - c_{\text{CL}_k}||_2^2 \to \min! \tag{3.8}$$

The number of clusters Nclust is fixed a priori and the cluster centers are initialized to random values.

Then, an iterative two-step algorithm, also referred to as **Lloyd's algorithm**, is applied [62].

First, each data point is assigned to the nearest cluster center. Then, in an update step, each cluster center is recomputed to its current mean. These steps are repeated until a stop criterion is met, e.g., the assignments do not change anymore or a threshold for the minimization task in Equation (3.8) is reached.

K-means always converges to a fixed point, but this is not necessarily the optimal solution. It can get stuck in a local minimum. Moreover, the K-means solution strongly depends on the initial condition. To overcome this drawback, K-means is often used with multiple restarts. Then, the best solution among them is chosen. **K-means++** [123], an extension of K-means, tries to improve the starting conditions by spreading the cluster centers away from each other, which leads to improvements in speed and accuracy of K-means.

The high computational cost of K-means due to the computation of pointwise distances can be decreased by efficient implementations, e.g., with kd-trees, see [62].

## 3.3.2  Principal Component Analysis Using Singular Value Decomposition

The **principal component analysis (PCA)** is a commonly used representative of linear dimension reduction methods.

PCA transforms the original observed variables (**A**) linearly into a set of derived variables, named the *principal components*, which are uncorrelated. This transformation is performed, so that as much variation present in the data set as possible is preserved. For this purpose, the covariance matrix of **A** is investigated.

The derivation of the principal components and a detailed discussion of PCA can be found in the textbooks [58, 53, 73].

A common and robust way to compute the principal components is by applying a singular value decomposition, which is explained in the following paragraph.

**Remark 3.12.** *Further linear dimension reduction methods related to PCA are, for example, **factor analysis** [44, 46] and **metric multi-dimensional scaling (MDS)** [69]. A comparison between PCA and factor analysis is given in [59, 58]. MDS decomposes the Gram matrix $\mathbf{A}^T\mathbf{A}$ instead of the covariance matrix $\mathbf{A}\mathbf{A}^T$. That is, $\mathbf{A}$ has not to be known explicitly. The computation of a singular value decomposition (SVD) of a matrix $\mathbf{A} \in \mathbb{R}^{\mathrm{Ndata}\times\mathrm{Nexp}}$ is generally based on the eigensystem of either $\mathbf{A}^T\mathbf{A}$ or $\mathbf{A}\mathbf{A}^T$, because the square roots of the eigenvalues of these matrices are the singular values of $\mathbf{A}$. Thus, the MDS method provides the same results as PCA, since a SVD of $\mathbf{A}$ can be used for the computation in both cases [73].*

**Singular Value Decomposition**

**Definition 3.13** (Singular Value Decomposition). *The **singular value decomposition (SVD)** of a matrix $\mathbf{A} \in \mathbb{R}^{\mathrm{Ndata}\times\mathrm{Nexp}}$, $\mathrm{Ndata} \geq \mathrm{Nexp}$ is defined as*

$$\mathbf{A} = \mathbf{U}\mathbf{\Lambda}\mathbf{V}^T, \tag{3.9}$$

*where $\mathbf{U} = (\mathbf{U}_1, \mathbf{U}_2, \ldots, \mathbf{U}_{\mathrm{Nexp}}) \in \mathbb{R}^{\mathrm{Ndata}\times\mathrm{Nexp}}$, $\mathbf{V}^T = (\mathbf{V}_1^T, \mathbf{V}_2^T, \ldots, \mathbf{V}_{\mathrm{Nexp}}^T) \in \mathbb{R}^{\mathrm{Nexp}\times\mathrm{Nexp}}$ are orthogonal matrices such that $\mathbf{U}^T\mathbf{U} = \mathbf{I}_{\mathrm{Nexp}}$ and $\mathbf{V}^T\mathbf{V} = \mathbf{V}\mathbf{V}^T = \mathbf{I}_{\mathrm{Nexp}}$. $\mathbf{\Lambda} \in \mathbb{R}^{\mathrm{Nexp}\times\mathrm{Nexp}}$ is a diagonal matrix with diagonal entries $\lambda_1 \geq \lambda_2 \geq \cdots \geq \lambda_{\mathrm{Nexp}} \geq 0$.*

**Remark 3.14.** *The SVD exists for every matrix in $\mathbb{R}^{\mathrm{Ndata}\times\mathrm{Nexp}}$ with $\mathrm{Ndata} \geq \mathrm{Nexp}$. The number of non-zero singular values determines the rank $r$ of $\mathbf{A}$, i.e., $\lambda_{r+1} = \lambda_{r+2} = \ldots = \lambda_{\mathrm{Nexp}} = 0$. The entries $\{\lambda_j\}$ are called the singular values of $\mathbf{A}$, the columns $\{\mathbf{U}_j\}$ and $\{\mathbf{V}_j\}$ are called left singular vectors and right singular vectors of $\mathbf{A}$, respectively.*

SVD transforms the data set into a new, derived geometric space [105], which is described by the product $\mathbf{\Lambda}\mathbf{V}^T$. As much variation as possible is expressed along each new axis independently from the variation expressed along the other, orthogonal axes.

**Remark 3.15.** *The magnitudes of values in different rows / columns of the matrix have to be comparable, when applying a PCA. Else the larger magnitudes will have a stronger influence on the resulting principal components than the smaller ones. Therefore, a normalization of the matrix may become necessary. A standard way to normalize the matrix is the normalization to zero mean, or to zero mean and standard unit [105].*

**Computation**

Classical methods for computing the SVD can be found, e.g., in [40, 41, 9, 91] and in standard textbooks of linear algebra. The complexity of computing the SVD of $\mathbf{A}$ based on the covariance matrix with $\mathrm{Ndata} \gg \mathrm{Nexp}$

is $\mathcal{O}(\text{Ndata}^2\text{Nexp})$ [105], which gets very large for large Ndata, resulting in a high computational time. Furthermore, it is usually required to store the entire matrix into memory. This may become difficult for very high dimensional matrices, like the database $\widetilde{\mathbf{M}}$ (cf. Section 2.1), which we will consider with the new methodology.

An efficient implementation of the SVD of a dense matrix with Ndata $\gg$ Nexp operates block by block so that it is not needed to store the entire matrix in memory [88, 20]. First, the Gram matrix $\mathbf{G} := \mathbf{A}^T\mathbf{A}$ is constructed by incrementing the entries of $\mathbf{G}$ for each row of $\mathbf{A}$. Then, the spectral decomposition of $\mathbf{G}$ is performed, for example, with a Jacobi algorithm [41]. Finally, the remaining $\mathbf{U}$-matrix is computed, leading to the following equations

$$\mathbf{G} = \mathbf{V}\mathbf{\Lambda}^2\mathbf{V}^T,$$
$$\mathbf{U} := \mathbf{A}\mathbf{V}\mathbf{\Lambda}^{-1}. \tag{3.10}$$

For memory requirements of this algorithm refer to [88].

**Dimension Reduction**

**Definition 3.16** (rank-$k$ approximation). *Let $\mathbf{A}$ be a matrix with SVD given by Equation (3.9), and $rank(\mathbf{A}) = r$. We define $\mathbf{A}^k$ as*

$$\mathbf{A}^k = \sum_{i=1}^{k} \mathbf{U}_i \lambda_i \mathbf{V}_i^T, \qquad \text{with } k < r. \tag{3.11}$$

$\mathbf{A}^k$ *is called **rank-$k$ approximation** of $\mathbf{A}$.*

With the above definition we can formulate the following well-known theorem, see [9].

**Theorem 3.17** (Eckart - Young [31]). *Let $\mathbf{A}$ be a matrix with given SVD (Equation (3.9)), and $rank(\mathbf{A}) = r$. Let $\mathbf{A}^k$ be a rank-$k$ approximation of $\mathbf{A}$. Furthermore, let $\mathbf{Z}$ be an arbitrary matrix in $\mathbb{R}^{\text{Ndata}\times\text{Nexp}}$. Then, it holds that*

$$\min_{rank(\mathbf{Z})=k} ||\mathbf{A} - \mathbf{Z}||_F^2 = ||\mathbf{A} - \mathbf{A}^k||_F^2 = \lambda_{k+1}^2 + \cdots + \lambda_r^2. \tag{3.12}$$

This means that $\mathbf{A}^k$ is the best rank-$k$ approximation of the matrix $\mathbf{A}$ in the Frobenius norm. The squared approximation error is exactly the sum of squares of the omitted singular values. This error can be controlled by specifying $k$, so that

$$\text{err}(k) < \varepsilon_{\text{SVD,abs}}, \tag{3.13}$$

for a given threshold $\varepsilon_{SVD,abs}$. Therefore, a linear dimension reduction with the SVD proceeds by setting the singular values $\lambda_{k+1}$ to $\lambda_r$ to zero.

In many cases the singular values rapidly decrease with $k$, so that only a few singular values need to be retained in order to obtain a small approximation error. Methods to choose $k$ range from ad hoc rules-of-thumb to formal tests of hypothesis and cross validation procedures. Commonly used graphical methods are difficult to interpret and cannot be automated easily. An overview and a critical discussion of such stopping rules is given in the textbooks [53, 58, 105] and references therein.

**Remark 3.18.** *We use the SVD for ensemble compression of the database containing simulation results within the PRO-CHAIN methodology in Section 5.3. This section also includes a further discussion on estimators of the approximation error.*

### 3.3.3 Nonlinear Dimension Reduction Methods

A state-of-the-art approach to deal with nonlinear, high dimensional data is provided by nonlinear dimension reduction (NLDR) methods. In particular, these methods aim at the construction of a low dimensional embedding of high dimensional data. Therewith, the curse of dimensionality should be overcome to some extend.

NLDR methods can be categorized into distance and topology preservation methods. However, all these methods have some drawbacks in common, most of all, their high computational effort. We summarize the most frequently used NLDR methods. After that, we discuss their limitations with regard to the applications considered in this work.

A detailed survey on the theoretical background, together with a comparison between several NLDR methods and some test examples, can be found, e.g., in [73], which provides the basis for this section.

Most NLDR methods consist of an optional clustering step and a nonlinear projection step. The clustering step, as described in Section 3.3.1, should be applied to overcome memory restrictions due to a large number of samples Ndata. These might occur, e.g., if interdistance matrices between the observed variables need to be stored. The projection step usually minimizes a certain cost function.

**Distance preservation** methods try to preserve pairwise distances in the observed data set. The distances cannot only be measured with spatial distances, but also with graph or geodesic distances. These measure the distance between two points along a manifold, instead of through the embedding space. Hence, the manifold has to be known a priori. Then, the shortest path is found, for example, with Dijkstra's algorithm [29]. Other distance measures can be build with kernel functions [102].

Two well-established spatial distance preservation methods are **Sammon's Nonlinear Mapping (NLM)** [100] and **Curvilinear Component Analysis (CCA)** [28]. Both methods minimize a similar cost function dependent on the spatial distances between the data points. The embedding of a out-of-sample point with NLM is difficult, since it is originally only a discrete algorithm. Some strategies for the embedding of out-of-sample points and extensions of the method are discussed in [73]. Moreover, NLM has originally been developed with the focus on mappings to 2- or 3-dimensional spaces. CCA includes an interpolation procedure to embed out-of-sample points. Compared with NLM, CCA is more flexible due to a weighting of the distances in the cost function. Nevertheless, the difficulties due to the optimization procedure still remain. Especially, the procedure can get stuck in a local minimum, or might need many iterations to converge.

Methods using graph distances include **geodesic NLM, Isomap, curvilinear distances** [72], and **kernel PCA** [102]. Isomap is a spectral NLDR method, which is closely related to PCA, but uses graph distances instead of Euclidean ones. The computational complexity is mainly determined by the computation of graph distances. This leads to $\mathcal{O}(\text{Ndata}^2 \log \text{Ndata})$ [73] using Dijkstra's algorithm.

**Kernel PCA** [102] first linearizes the underlying manifold with a mapping $\phi$ in a possibly even higher dimensional space. This mapping $\phi$ is never evaluated explicitly, but only over kernel functions $\kappa$ acting as scalar products. The pairwise scalar products of the mapped data are computed, and stored in a matrix $\Phi$. Then, a spectral decomposition is applied to this symmetric matrix, analogously to the PCA procedure. Examples of valid kernel functions are (Gaussian) radial basis functions, polynomial kernels, or sigmoid kernels. Memory and time complexities remain almost the same as for metric MDS. The main difficulty is to choose an appropriate kernel. This is due to the fact that kernel PCA is not motivated geometrically, and the kernel functions are hard to interpret geometrically. Therefore, Kernel PCA is not used much in dimensionality reduction, but rather in support vector machines, which can be used for classification and regression.

**Topology preservation** methods investigate local neighborhood relationships of the data set, like inequalities, ranks, and angles. A difficult task is the characterization of the manifold, which is either predefined discretely, or data driven. In the latter case, the shape of the manifold can be modified within the algorithm.

Well-established topology preservation methods include the **Kohonen's Self Organizing Maps (SOM)** [67], **Local Linear Embedding (LLE)** [101] and **Laplacian Eigenmaps (LE)** [7]. The SOM method has been especially developed for the visualization of high dimensional data. This is one of the reasons why SOM is usually restricted to low, mainly 1- or 2-

dimensional embedding spaces. Furthermore, the structure of the discrete embedding space is defined a priori, usually hexagonal meshes are used. The embedding is done by an incremental learning algorithm. **Locally Linear Embedding (LLE)** [101] is a more recently developed spectral method based on *conformal mappings*, that is, local angles are preserved. Each point is reconstructed locally by a linear combination of its neighbors. Then, a sparse eigenvalue problem has to be solved and the smallest eigenvalues are taken. Therefore, a specialized eigenvalue decomposition procedure has to be used. **Laplacian eigenmaps (LE)** [7] is a closely related spectral method, which tries to overcome some drawbacks of LLE. LE is based on the constrained minimization of local distances. In the cost functions the distances are weighted with a kernel function related to the adjacency matrix, which provides the required neighborhood relationships. A common choice is the Gaussian kernel, see, e.g., [73]. The embedding is then provided by the smallest eigenvectors of the graph Laplacian. The spectral decomposition indicates that it performs better for data clustering.

**Remark 3.19.** *In the applications considered in this work, often only few samples (a few tens up to few hundreds) are available and the number of parameters is quite high (5 to 30). Moreover, the data represents criteria on an entire mesh for several timesteps, resulting in large dimensional vectors (dimension of several thousands up to millions).*

The above summary of NLDR methods shows that some drawbacks limit their efficient application to industrial problems. In detail, in the class of spectral NLDR methods, only the choice of the distance or kernel matrix realizes the nonlinearity, then a linear projection is applied to the so-constructed matrix. Most non-spectral NLDR methods are based on (stochastic) gradient descent optimization procedures, which can get stuck in a local minimum, or need many iterations to converge. Therefore, dependent on the initialization, different embeddings can be obtained.

Furthermore, NLDR methods often suffer from a high computational effort, e.g., quadratic in the number of data and memory requirements. The comparison between several methods in [73] shows that most of them already require a large number of samples (several hundreds) to identify the method parameters. On the other hand, if huge data sets (several thousand samples) are available, a clustering (cf. Section 3.3.1) has to be performed in advance due to the enormous computational effort otherwise. Additionally, most NLDR methods are not recommended for very high data dimensionality (Npar > 50). It has been shown, e.g., in [73], that most NLDR methods may perform poor when the intrinsic dimension is higher than 5. In conclusion, these drawbacks make the application of several NLDR methods difficult for our purpose.

Due to the high computational complexity of most methods and their difficulty to embed into higher dimensional spaces, we do not follow these methods further. Instead, we combine a parameter classification procedure (developed in Chapter 4) with a nonlinear RBF-metamodeling approach accelerated by a linear SVD (derived in Chapter 5 and Chapter 6).

## 3.4   Metamodels

Metamodels are commonly used to approximate computationally expensive functions in order to analyze the dependencies between parameters and the response. There are many different approximation approaches, cf. Section 2.2.2. In the following, we focus on metamodels with RBFs due to several advantages. First, RBF metamodeling does not require any specific assumption on the set of sampling points, therefore, it is well suited for scattered data. Second, polynomials are interpolated exactly, up to a certain degree, due to a possible polynomial detrending. Furthermore, learning procedures are not necessary [88, 60] compared with other approximation methods, like artificial neural networks. This is particularly because the function parameter $\gamma$, introduced below, does not influence the interpolation results except for smoothness. Finally, a uniform distribution of the sampling points in the parameter space provides fast convergence [13].

### Metamodels Based on Radial Basis Functions

Let $b$ denote a representative of a function $b\colon \mathbb{R}^{\mathrm{Npar}} \to \mathbb{R}^{\mathrm{Ncrit}}$ and let $\tilde{b}$ denote its approximation. In the following, we consider Ncrit $= 1$ without loss of generality.

**Definition 3.20** (Radial basis function (RBF)). *A RBF $\phi : \mathbb{R} \to \mathbb{R}$ is a real-valued function whose values depend only on the distance to the origin,* $\phi(x) = \phi(||x||_2)$.

Table 3.1 shows some commonly used types of RBFs for $r = ||x||_2$, for details refer to [13, 124]. The constant function parameter $\gamma > 0$ controls the smoothness of the function.

A RBF metamodel [13] is a finite linear combination of RBFs,

$$\tilde{b}(\mathbf{P}) = \sum_{l=1}^{\mathrm{Nexp}} \phi(||\mathbf{P} - \mathbf{P}_l||_2)c_l, \qquad (3.14)$$

with coefficients $c_l \in \mathbb{R}$, which are determined so that Equation (2.22) holds, i.e.,

$$\tilde{b}(\mathbf{P}_k) = \sum_{l=1}^{\mathrm{Nexp}} \phi(||\mathbf{P}_k - \mathbf{P}_l||_2)c_l, \quad k = 1, \ldots, \mathrm{Nexp}. \qquad (3.15)$$

| Name | Function |
|------|----------|
| Linear | $\phi(r) = r$ |
| Thin-plate spline | $\phi(r) = r^2 \log r$ |
| Gaussian | $\phi(r) = \exp(-\gamma r^2), \gamma > 0$ |
| Multiquadric | $\phi(r) = \sqrt{r^2 + \gamma^2}, \gamma > 0$ |
| Triharmonic spline | $\phi(r) = r^3$ |
| Inverse multiquadric | $\phi(r) = (r^2 + \gamma^2)^{-\frac{1}{2}}$ |

Table 3.1: Commonly used types of radial basis functions.

We define an interpolation matrix $\mathbf{\Phi} = (\phi(||\mathbf{P}_k - \mathbf{P}_l||_2))_{k,l=1,\ldots,\text{Nexp}}$. Therewith, Equation (3.15) becomes $\tilde{\mathbf{b}} = \mathbf{\Phi}\mathbf{c}$ and $\mathbf{c} = \mathbf{\Phi}^{-1}\tilde{\mathbf{b}}$. Note that the interpolation matrix is always nonsingular for the multiquadric function. Details of the invertibility of the interpolation matrix can be found in [13, 124]. By inverting the interpolation matrix, the result $\tilde{b}$ can be written as weighted sum [87]

$$\tilde{b}(\mathbf{P}) = \sum_{k=1}^{\text{Nexp}} w_k(\mathbf{P})b(\mathbf{P}_k), \tag{3.16}$$

with

$$w_k(\mathbf{P}) = \sum_{l=1}^{\text{Nexp}} \Phi_{kl}^{-1}\phi(||\mathbf{P} - \mathbf{P}_l||_2), \tag{3.17}$$

where $\Phi_{kl}^{-1}$ are the entries of $\mathbf{\Phi}^{-1}$. This metamodel can be extended by polynomial detrending, that is, by adding a polynomial part $\Psi_{\text{Dd}}$ of degree at most Dd

$$\tilde{b}(\mathbf{P}) = \sum_{l=1}^{\text{Nexp}} \phi(||\mathbf{P} - \mathbf{P}_l||_2)c_l + \Psi_{\text{Dd}}(\mathbf{P}). \tag{3.18}$$

Polynomial detrending improves the precision of interpolation and leads to an exact representation of polynomials up to the degree Dd, for more details see, for example, [13].

To avoid the degeneracy of the interpolation matrix by the extra degrees of freedom using the polynomial part, the additional condition

$$\sum_{l=1}^{\text{Nexp}} \mathbf{P}_l^\alpha c_l = 0, |\alpha| \leq \text{Dd}, \tag{3.19}$$

where $\alpha$ denotes a multiindex, has to be fulfilled. Specifically, the number of monomials building a basis of the space of polynomials of degree at most Dd in Npar unknowns has to be smaller than the number of sampling points, i.e.,

$$\sum_{j=0}^{\mathrm{Dd}} \binom{\mathrm{Npar}-1+j}{j} < \mathrm{Nexp} \qquad (3.20)$$

holds.

The RBF metamodel introduced also allows the efficient model tolerance prediction. In detail, in combination with the response also its precision can be predicted. The model tolerance is based on **cross validation**, which measures the sensitivity of the metamodel on the removal of a sampling point. The cross validation proceeds as follows. An arbitrary sampling point $\mathbf{P}_k$ is removed from the set of sampling points, then a new metamodel $\tilde{b}(\mathbf{P}^{(k)})$ is constructed with the remaining $\mathrm{Nexp}-1$ points. The removed experiment is predicted with the metamodel and compared with the existing value of the sampling point. The so computed difference in this sampling point is used as a predictor of the metamodel error. In detail, a high tolerance indicates a high sensitivity on the removal of sampling points. Thus, also a higher error may arise. This procedure can be carried out for all Nexp sampling points. The tolerance in out-of-sample points can be interpolated as well. For a RBF metamodel with detrending, the cross validation procedure does not require the explicit construction of additional metamodels. Instead, a direct formula can be derived analytically.

**Theorem 3.21** ([110, 87, 21]). *Let a RBF metamodel with polynomial detrending be given by Equation (3.18). Then the model quality in an arbitrary point can be predicted by the tolerance* tol($\mathbf{P}$)

$$\mathrm{tol}(\mathbf{P}) = \tilde{b}(\mathbf{P}^{(k)}) - \tilde{b}(\mathbf{P}) = -\frac{w_k(\mathbf{P})c_k}{\Phi_{kk}^{-1}}, \qquad (3.21)$$

*where* $\tilde{b}(\mathbf{P}^{(k)})$ *denotes the cross validated metamodel in which the k-th sample is removed.*

A proof and further details can be found in my diploma thesis [110]. This procedure can be carried out over all sampling points. A total model tolerance in an arbitrary point is derived by accumulating the tolerances (Equation (3.21)) in a proper norm, see [110]. A major advantage of cross validation is that the interpolation quality can be predicted avoiding additional simulations runs.

The number of samplings required to get a high quality metamodel depends on the dimension of the parameter space and the degree of the polynomial detrending used. In order to apply a detrending with $\mathrm{Dd} = 2$, the minimal number of samples can be computed by

$$\mathrm{Nexp_{min}} = C(2 + \mathrm{Npar} + \mathrm{Npar}(\mathrm{Npar}+1)/2), \qquad (3.22)$$

fulfilling the condition given by Equation (3.20), cf. [110, 34]. Numerical experiments have shown that the constant $C$ should be chosen to be an integer in $[1, 5]$. Moreover, they have shown that it is usually well suited to use $C \geq 3$ and

$$\text{Nexp} = \max\left(20, \text{Nexp}_{\min}\right) \tag{3.23}$$

in nonlinear applications.

## 3.5 Stochastic Finite Element Methods

The conventional approach to solve a **stochastic partial differential equation (sPDE)**, that is, a PDE with random input variables, is a MC method. Recently, new stochastic finite element methods have been developed, like the **stochastic Galerkin** and the **stochastic collocation** method, see, for example, the review paper [126]. These methods enable the computation of high-order stochastic solutions with less computational effort than MC methods by transforming the sPDE into a set of coupled deterministic PDEs in the Galerkin case, or a set of uncoupled deterministic PDEs in the collocation case, respectively. However, both methods suffer from the curse of dimensionality. While the collocation method is a sampling based solution method as MC, the Galerkin approach is a direct and, thus, sampling-free solution approach.

**Remark 3.22.** *In this thesis, we focus on sampling methods, since the underlying system of PDEs is usually not given explicitly in the applications considered. Hence, we discuss only the collocation method in detail, as an alternative sampling based interpolation method. A numerical comparison between the collocation method and the metamodeling technique used within the PRO-CHAIN methodology is given in Section 7.1.*

Let $\mathcal{D}$ be a spatial domain with coordinates $\mathbf{X}$. A sPDE is defined as

$$\mathcal{L}(\mathbf{X}, \mathbf{P}, u) = 0, \quad \text{in } \mathcal{D}, \tag{3.24}$$
$$\mathcal{B}(\mathbf{X}, \mathbf{P}, u) = 0, \quad \text{on } \partial\mathcal{D},$$

where $\mathcal{L}$ is a general differential operator and $\mathcal{B}$ is a boundary operator properly specified. Assuming the Npar parameters $P^j$ are independent random variables with probability density function $f_{P^j}$ with image $\Gamma_j$, their joint probability density equals $f_{\mathbf{P}} = \prod_{j=1}^{\text{Npar}} f_{P^j}$ with the support $\Gamma = \prod_{j=1}^{\text{Npar}} \Gamma_j$. Then, we can investigate the finite dimensional random space $\Gamma$ and search for a solution $u$ of the sPDE valid for all $\mathbf{P} \in \Gamma$, cf. [126].

Thus, a main issue is to approximate the sources of randomness by a finite number of independent random variables [126]. This *finite dimensional noise assumption* [4] is satisfied, e.g., by the formulation of a truncated **Karhunen-Loève (KL)** expansion. The KL expansion is based on the spectral decomposition of the covariance function of a process $\alpha(\mathbf{X}, \omega)$, that is,

$$\alpha(\mathbf{X}, \omega) = \mathbf{E}[\alpha(\mathbf{X})] + \sum_{j=1}^{\text{Npar}} P^j(\omega)\sqrt{\lambda_j}\phi_j(\mathbf{X}), \qquad (3.25)$$

where $\lambda_j$ and $\phi_j$ are eigenvalues and eigenvectors of the covariance function [38]. The number of parameters is specified so that the error due to the truncation of the expansion is minimized.

Considering the stochastic solution $u$ of the sPDE, the covariance function is not known. The **generalized polynomial chaos (gPC)** expansion [38, 126, 128] approximates a random function with orthogonal polynomials of the random variables

$$u(\mathbf{X}, \mathbf{P}) \approx \sum_{k=1}^{\text{Nexp}} u(\mathbf{X}, \mathbf{P}_k)\Psi_k(\mathbf{P}). \qquad (3.26)$$

A complete orthogonal basis consists of $\text{Nexp} = \binom{\text{Npar}+\Psi_D}{\text{Npar}}$ basis functions, where $\Psi_D$ denotes the polynomial order [126]. The orthogonal polynomials chosen depend on the joint probability distribution of the random variables. For example, Hermite polynomials are used in the case of Gaussian distributed random variables resulting in the original *Wiener polynomial chaos* expansion. A possible approach to minimize the truncation error and to compute the stochastic solution is to apply a Galerkin projection.

In the following, a brief introduction of the Galerkin approach is given in Section 3.5.1. Then, the collocation method is explained in Section 3.5.2.

## 3.5.1 Stochastic Galerkin Approach

The **stochastic Galerkin** approach [38, 126] uses the gPC expansion given by Equation (3.26). Then, the expansion coefficients $u_k(\mathbf{X})$ are obtained by a Galerkin projection of the differential equation onto the polynomial basis functions. The weak formulation of Equation (3.24) has to be satisfied for all basis functions, i.e.,

$$\int \mathcal{L}\left(\mathbf{X}, \mathbf{P}, \sum_{k=1}^{\text{Nexp}} u(\mathbf{X}, \mathbf{P}_k)\Psi_k(\mathbf{P})\right)\Psi_l(\mathbf{P})f_\mathbf{P}\, d\mathbf{P} = 0, \quad \text{in } \mathcal{D}, \qquad (3.27)$$

$$\int \mathcal{B}\left(\mathbf{X}, \mathbf{P}, \sum_{k=1}^{\text{Nexp}} u(\mathbf{X}, \mathbf{P}_k)\Psi_k(\mathbf{P})\right)\Psi_l(\mathbf{P})f_\mathbf{P}\, d\mathbf{P} = 0, \quad \text{on } \partial\mathcal{D}.$$

This results in a coupled system of deterministic PDEs and standard numerical techniques can be applied. The main advantage of the stochastic Galerkin method is its fast convergence when the solution depends smoothly on the random variables [126]. In this case, a high-order stochastic solution can be found with a limited number of equations. However, the solution of the integral Equations (3.27) can become nontrivial [126] requiring highly efficient and robust parallel solvers. For a discussion on the limitations and advantages of the Galerkin approach refer to [126] and references therein.

### 3.5.2 Stochastic Collocation Method

The following summary of the stochastic collocation method is based on the publication [115]. As already stated, the stochastic collocation method is a sampling based solution method for sPDEs, which leads to uncoupled deterministic problems. An extended overview of this method can be found, for example, in [4, 127]. The solution $u$ is represented discretely by an expansion with multivariate Lagrange polynomials $l_k(\mathbf{P})$ satisfying the property $l_k(\mathbf{P}_m) = \delta_{k,m}$, with $\delta_{k,m}$ the Kronecker delta. Hence, the solution is given by

$$u(\mathbf{X}, \mathbf{P}) \approx \sum_{k=1}^{\text{Nexp}} u(\mathbf{X}, \mathbf{P}_k) l_k(\mathbf{P}). \tag{3.28}$$

By requiring that Equation (3.24) holds in each collocation point $\mathbf{P}_k, k = 1, \ldots, \text{Nexp}$, we obtain a set of Nexp decoupled deterministic PDEs,

$$\forall k = 1, \ldots, \text{Nexp} :$$
$$\mathcal{L}(\mathbf{X}, u(\mathbf{X}, \mathbf{P}_k), \mathbf{P}_k) = 0, \quad \text{in } \mathcal{D}, \tag{3.29}$$
$$\mathcal{B}(\mathbf{X}, u(\mathbf{X}, \mathbf{P}_k), \mathbf{P}_k) = 0, \quad \text{on } \partial\mathcal{D}.$$

The construction of the set of multidimensional collocation points is essential for the computational effort of the method, since Nexp deterministic problems have to be solved. Hence, the number of collocation points required for an accurate solution should be minimized. The collocation points are typically constructed as a sparse grid of Gauss or Clenshaw-Curtis points, cf. Section 3.1. Therewith, the stochastic collocation method achieves a fast convergence rate [127, 126].

Given the solution by Equation (3.28), the statistics of the solution can be evaluated. By choosing the collocation points to be a cubature point set [126, 127], for example, the mean can be estimated as

$$\mathbf{E}[u(\mathbf{X}, \mathbf{P})] = \sum_{k=1}^{\text{Nexp}} u(\mathbf{X}, \mathbf{P}_k) w_k, \tag{3.30}$$

where $w_k$, $k = 1, \ldots, \text{Nexp}$, are the cubature weights corresponding to the cubature points $\mathbf{P}_k$ [76, 115]. With sampling the stochastic solution, given by Equation (3.28), in the parameters by a MC method, other statistics, e.g., quantiles, can be evaluated straightforwardly.

A major advantage of the stochastic collocation approach is that it permits a direct reuse of deterministic solvers. The number of collocation points still grows fast in the case of a large number of random parameters, which limits the application of stochastic collocation methods. Current research aims at reducing the required number of simulations further, for example, by taking the anisotropy of the problem into account [115, 89, 11].

# Chapter 4

# Parameter Classification Using Sensitivity Analysis

Many different parameters are involved in each processing step. Each parameter may be subject to scatter. The parameter classification procedure is the first essential component in the PRO-CHAIN methodology. It quantifies the amount of influence of these parameter variations on the criteria considered. The classification results are a necessary basis for the minimization of the computational effort of subsequent analysis steps in order to enable the local transfer of all relevant scatter information to the next processing step.

The parameters are classified with respect to the influence of their variations on a criterion considered. For this purpose, we use a local sensitivity analysis. A sensitivity analysis aims at sorting the parameters, such as material properties, forces, and friction, by their decreasing amount of influence on the criterion under consideration, e.g., sheet thickness, strains, and stresses.

We have discussed state-of-the-art parameter sensitivity methods in Section 3.2. First, we have pointed out that the parameter sensitivity is usually measured globally. Hence, local effects cannot be resolved. However, we have observed (cf. Section 2.3 and Chapter 7) that physical effects often influence the criterion only locally for the applications considered. Moreover, several locally separated phenomena can be observed, which do not affect each other. Therefore, it is important to analyze these local effects.

Second, the discussion has shown that global variance-based methods, as the Sobol indices, are highly computationally expensive. Hence, these methods would only be practicable in industrial applications, if they are applied to metamodels, which approximate the underlying problem. Thus, the sensitivities computed depend on the quality of the metamodel.

To address these drawbacks, we develop a **parameter classification procedure** using a local parameter sensitivity analysis within the new PRO-CHAIN methodology that aims at reducing the dimension of the parameter space, and at locating locally interesting areas of the simulated component, where possible defects, like fractures, may occur. The approach proposed has the following three major advantages.

First, the parameter classification is **fully automated** ensuring that parameters with a similar proportion of influence are assigned to the same importance class.

Second, we introduce a **local sensitivity measure** in each mesh node in order to detect important local effects within the simulation process. Furthermore, this provides a local estimate of the average prediction accuracy of a forecast model created with a correspondingly reduced set of parameters.

Third, the parameter classification minimizes the required number of simulation runs. Thus, it is **efficiently applicable** to complex industrial problems without the need of a metamodel. In conclusion, metamodels used as forecast models can be created on the reduced set of parameters, which highly reduces the number of simulation runs required.

The results of the parameter classification are directly used in the subsequent PRO-CHAIN component, namely the dimension reduction and ensemble compression of the database, as explained in Chapter 5.

The developed parameter classification procedure is derived in **Section 4.1**. We investigate a clustering of the mesh nodes based on the local nonlinearity measure in **Section 4.2** in order to identify regions with similar parameter influences. This decomposition is used in subsequent analysis steps in order to reduce the computational effort further, and to increase the accuracy, by applying different methods in local regions of interest. The computational effort and memory requirements of the parameter classification procedure is investigated in **Section 4.3**. A discussion of the classification procedure including a comparison with state-of-the-art sensitivity analysis methods is given in **Section 4.4**.


## 4.1   Importance and Nonlinearity Classes

The parameter classification procedure assigns each parameter to three classes based on the result of the sensitivity analysis.

A parameter is said to be **important**, if the variation of this parameter by the amount of its standard deviation causes also a relatively high variation of the criterion compared with the other parameter influences.

The influence of a parameter on a criterion is said to be **linear**, if the change in the criterion can be described by a linear function applied to this criterion. In this case, the first partial derivatives are sufficient to describe the amount of change. A parameter is said to be **nonlinear** if second partial derivatives or higher order derivatives have a substantial influence.

On the one hand, each parameter is assigned to a **linear importance class**, which indicates the relative importance of this parameter on the criterion. Parameters possessing no or only a small influence on the criterion are negligible and can be fixed to their mean value. Thus, these parameters are excluded from a further analysis. This leads to a reduction of the dimension of the parameter space. In contrast, parameters which have a strong influence on the criterion should be investigated in more detail in order to obtain an appropriate prediction accuracy, for example, in a subsequent forecast model, see Chapter 6.

On the other hand, the parameters are characterized according to a measure of nonlinearity for an impact of their variations on the criterion. These classes are called **nonlinearity classes** and investigate the nature of the behavior of parameter variations on the criterion. If the parameter influences the criterion in a linear way, a linear interpolation method with two or three simulations will be sufficient to get a low interpolation error when predicting the behavior of a new design. Otherwise, if the behavior is nonlinear, the design of experiments (DoE) needs to be extended to allow a more sophisticated interpolation method in order to approximate the criterion appropriately. That is, the distribution of the parameter and the resulting distribution of the criterion have to be described in more detail in oder to get an accurate forecasting.

Finally, each parameter is assigned to a **total importance class**, which summarizes all effects. These total importance classes consider the overall influence on the criterion given by the sum of the linear and nonlinear influences. The usage of the total importance classes simplifies an appropriate parameter space reduction.

The derived linear importance classes and appropriate measures are explained in detail in Section 4.1.1, whereas the nonlinearity classes are developed in Section 4.1.2. The total importance classes are introduced in Section 4.1.3. In Section 4.1.4, the parameter classification is applied to several test problems in order to illustrate the procedure and to show the differences among the classification measures.

**Remark 4.1.** *Parts of the following parameter classification procedure have already been published in [115].*

## 4.1.1    Linear Importance Classes

In the parameter classification procedure, each parameter is characterized by its relative importance to the criterion considered compared with the other parameters at hand. For this purpose, each parameter is assigned to a linear importance class. The measure of importance is based on a local sensitivity analysis. Thus, the change of the criterion due to a change in the parameters is investigated in each mesh node. More precisely, motivated by the Taylor expansion, the importance measure is based on the approximation of partial derivatives with respect to the parameters using a finite difference scheme, cf. Section 3.2.1. Note that the linear importance classes only take the first derivative into account. Thus, only changes in the criterion caused by a linear function applied can be detected and quantified.

**Remark 4.2.** *As already mentioned, we can suppose that the nominal parameter set and corresponding solution is known in the applications considered in this work. We aim at finding a robust solution around this nominal solution with help of a sensitivity analysis. A sensitivity analysis is performed in order to investigate the robustness of this solution and to transfer the influences of parameter variations from one processing step to the next step. Therefore, a local consideration of parameter variations and the application of local sensitivity methods is suitable in this case.*

The parameter classification is based on a database containing simulation results for a specific criterion and proceeds as follows. It starts by constructing a database using a star-point DoE. The DoE is applied as explained in Section 3.1 and $2\text{Npar} + 1$ simulation runs are performed accordingly. With these simulation results we set up a database $\mathbf{M}$ (2.1) for each criterion $Y^j, j = 1, \ldots \text{Ncrit}$, to be considered, such as sheet thickness, effective plastic strain (EPS), and damage information.

**Remark 4.3.** *The same simulation runs can be used for the parameter classification of all criteria. The extracted simulation results corresponding to a specific criterion are stored in separate databases.*

**Example 4.4.** *Considering the criteria vector $\mathbf{Y} = [Y^1, Y^2]^T = [EPS, t]^T$, we construct a database containing results of EPS and another database containing results of t.*

**Remark 4.5.** *In the following, the criterion $Y^j, j \in \{1, \ldots, \text{Ncrit}\}$, is used as arbitrary representative for one of the criteria considered. $\mathbf{g}$ denotes the corresponding function, and $\mathbf{M}$ the corresponding database constructed, cf. Section 2.1. The individual PRO-CHAIN components are applied for each criterion separately, unless stated otherwise.*

In the following, we suppose that the first two derivatives of all functions $g_i, i = 1, \ldots, \text{Nnodes}$ exist and are continuous, i.e., $g_i \in \mathcal{C}^2$. Then, the first partial derivatives of the function $\mathbf{g}$ are approximated with a central finite differences (FD) method using the database $\mathbf{M}$. This results in a second order FD approximation of the *Jacobian matrix* in each mesh node $N_i$ given as $\mathbf{J} = (J_{ij})_{i=1,\ldots,\text{Nnodes},j=1,\ldots,\text{Npar}}$ with entries

$$J_{ij} := \frac{\partial g_i(\mathbf{P})}{\partial P^j} = \frac{g_i(\mathbf{P}^{j+1}) - g_i(\mathbf{P}^{j-1})}{2\sigma_{P^j}} + \mathcal{O}(\sigma_{P^j}^2), \tag{4.1}$$

where all parameters except $P^j$ are held fixed, see Equation (2.5). $\sigma_{P^j}$ denotes the variation range of the parameter $P^j$ as defined in Section 2.

We define a matrix $\mathbf{S1} = (S1_{ij})_{i=1,\ldots,\text{Nnodes},j=1,\ldots,\text{Npar}}$ [12] with entries

$$S1_{ij} := \sigma_{P^j} J_{ij}, \tag{4.2}$$

called **local sensitivity** matrix, to measure the local linear influence of parameter variations in each mesh node. The local sensitivity is given by an approximation of the first term of the Taylor expansion reflecting only the changes due to linear influences. $S1_{ij}$ can be interpreted as the variation in mesh node $N_i$ of the criterion considered due to the variation of parameter $P^j$ by $\sigma_{P^j}$. In contrast to the Jacobian values, the local sensitivity matrix allows a direct comparison of sensitivity values for different parameters because all entries are measured in the same unit as the criterion considered.

**Remark 4.6.** *The use of a second order FD method instead of a first order method does not cause any additional effort because the simulations with parameter sets $\mathbf{P}^{j-1}$ and $\mathbf{P}^{j+1}$ have to be performed anyway for the classification into nonlinearity classes, cf. Section 4.1.2. The number of simulation runs is kept at a minimum, since $2\text{Npar}+1$ is the minimal number of simulation runs required to compute all sensitivity measures needed for the overall parameter classification.*

Therewith, we have computed a measure to compare parameter influences locally. This gives a valuable insight into the simulation process by detecting important local influences and corresponding domains of the entire mesh. Nevertheless, the simulation run has to be always performed on the entire component to be physically meaningful.

Due to this fact, we define an accumulated sensitivity measure based on the local one in order to decide which parameter variations should be considered further. The accumulated sensitivity measure is required to allow an overall ranking of parameters w.r.t. their influence on the criterion considered. The accumulated sensitivity measure is defined as expected value of the absolute local sensitivity measures. An accumulated Jacobian matrix can be expressed as

$$J^j := \frac{1}{\text{Nnodes}} \sum_{i=1}^{\text{Nnodes}} |J_{ij}|, \quad j = 1, \ldots, \text{Npar}. \tag{4.3}$$

Consequently, the **accumulated sensitivity measure** of parameter $P^j$ is defined as

$$S1^j := \sigma_{P^j} J^j. \tag{4.4}$$

Note that we sum absolute values in order to avoid that positive and negative influences add up to zero.

**Remark 4.7.** *The accumulated sensitivity measure differs from a global sensitivity measure because the sensitivities are accumulated over the mesh nodes. That is, local sensitivities around a fixed parameter vector are still investigated. Global sensitivity measures investigate the entire parameter space. This can be compared with an accumulation over the parameter space.*

**Remark 4.8.** *We have used the $L_2$-norm over all mesh nodes as accumulated sensitivity measure in [115]. The usage of the expected value (Equation (4.3)) improves the interpretation of the sensitivity values derived, since the measure is in the same unit as the criterion considered, cf. [12].*

The vector of accumulated measures, given by Equation (4.4), is normalized to $[0, 1]$ by dividing each entry by the sum over all accumulated sensitivity measures, that is,

$$\widetilde{S1}^j = \frac{S1^j}{\sum_{j=1,\ldots,\text{Npar}} S1^j}, \quad j = 1, \ldots, \text{Npar}. \tag{4.5}$$

$\widetilde{S1}^j$ represents the proportion of the overall variation according to parameter $P^j$. Note, if $J^j$ is almost zero, e.g., $J^j < 1 \times 10^{-5}$, $\widetilde{S1}^j$ is set to zero. That is, very small influences, which are negligible, are set to zero in order to avoid a lot of additional linear importance classes with almost no influence.

Finally, the parameters are ranked by sorting the normalized accumulated sensitivity measures (Equation (4.5)) in descending order. To be more specific, each parameter $P^j, i = 1, \ldots, \text{Npar}$ is classified into a **linear importance class** $\text{LIClass}_k, k = 1, 2, \ldots$ by the following procedure that ensures that parameters having a similar proportion of influence on the criterion are assigned to the same linear importance class.

First, each parameter with a normalized accumulated sensitivity measure equal or greater than the average percentage of influence is assigned to the first linear importance class $\text{LIClass}_1$, i.e.,

$$\forall j \in \{1, \ldots, \text{Npar}\} : \widetilde{S1}^j \geq \frac{1}{\text{Npar}} \Rightarrow P^j \in \text{LIClass}_1. \tag{4.6}$$

Hence, the first importance class contains the most influential parameters. Analogously, we consider the remaining, not classified parameters, that is,

$$\forall j \in \{1, \ldots, \text{Npar}\}, P^j \notin \text{LIClass}_1 :$$

$$\frac{S1^j}{\sum_{j=1,\ldots,\text{Npar},P^j \notin \text{LIClass}_1} S1^j} \geq \frac{1}{\text{Npar} - \#\text{LIClass}_1} \tag{4.7}$$

$$\Rightarrow P^j \in \text{LIClass}_2,$$

where $\#\text{LIClass}_1$ denotes the cardinality of $\text{LIClass}_1$, which is the number of elements of the set, as defined in Section 2.1. Note that the normalization of the accumulated sensitivity measures has to be recomputed with the remaining parameters only. Finally, this procedure is iteratively repeated until all parameters are assigned to an importance class. Thus, linear importance classes with a higher index $k$ contain parameters with fewer linear influence. The total number of importance classes is determined by the similarity structure of the accumulated parameter influences. In the worst case, the assignment procedure terminates after Npar iterations. In this case, each class contains exactly one parameter.

**Example 4.9.** *Assume that all parameters have equal influence on the criterion. In this case, the percentage of influence will be $1/\text{Npar}$. Then, all parameters are assigned to the first linear importance class $\text{LIClass}_1$ according to Equation (4.6).*

This procedure ensures that all parameters with a similar proportion of influence are included in the same importance class in order to facilitate the selection of parameters remaining in the analysis. The selection of parameters and the corresponding reduction of the dimension of the parameter space is described in Section 5.1.

### 4.1.2 Nonlinearity Classes

The Jacobian matrix only shows the magnitude of the linear influence of each parameter on the criterion.

Hence, additionally, an indicator for the nonlinear influence of the parameters on the criterion is computed. This indicator uses the diagonal part of the Hessian matrix as a weak approximation of the full Hessian matrix. The full Hessian matrix is not computed in order to minimize the computational effort.

The Hessian matrix $\mathbf{H} = (H_{jl})_{j,l=1,\ldots,\text{Npar}}$ is called *weakly diagonally dominant*, if

$$r_l := \sum_{j=1,j\neq l}^{\text{Npar}} |H_{jl}| \leq |H_{ll}|, \quad \forall l = 1, \ldots, \text{Npar} \tag{4.8}$$

holds. A Gerschgorin disc $\mathcal{G}_l(\mathbf{H})$ for the square matrix $\mathbf{H}$ is defined as the closed disc with radius $r_l$ around $H_{ll}$, that is,

$$\mathcal{G}_l(\mathbf{H}) := \{\alpha \in \mathbb{C} \ : \ |\alpha - H_{ll}| \leq \sum_{j=1,j\neq l}^{\mathrm{Npar}} |H_{jl}|\}. \qquad (4.9)$$

The Gerschgorin theorem states that every eigenvalue of $\mathbf{H}$ lies in one of the Gerschgorin discs $\mathcal{G}_l$. With Equation (4.8), it holds for each eigenvalue $\alpha$ of $\mathbf{H}$

$$|\alpha - H_{ll}| \leq \sum_{j=1,j\neq l}^{\mathrm{Npar}} |H_{jl}| \leq |H_{ll}|. \qquad (4.10)$$

Therefore, the diagonal part of the Hessian matrix is a good weak indicator of the full Hessian matrix in an appropriate norm as long as the Hessian matrix is weakly diagonally dominant.

Analogous to the Jacobian matrix, the diagonal part of the Hessian matrix $\mathbf{H}$ is approximated in each mesh node $N_i$ with a second order FD method. The matrix whose rows are the diagonal entries of the Hessian matrix in the corresponding mesh node is defined as $\mathbf{diag}H$. Its entries $\mathrm{diag}H_{ij}, i = 1, \ldots, \mathrm{Nnodes}, j = 1, \ldots, \mathrm{Npar}$ are given by

$$\mathrm{diag}H_{ij} := \frac{\partial^2 g_i(\mathbf{P})}{\partial(P^j)^2} \approx \frac{g_i(\mathbf{P}^{j+1}) - 2g_i(\mathbf{P}) + g_i(\mathbf{P}^{j-1})}{\sigma_{P^j}^2}. \qquad (4.11)$$

We define a matrix $\mathbf{S2} = (S2_{ij})_{i=1,\ldots,\mathrm{Nnodes},j=1,\ldots,\mathrm{Npar}}$ with entries

$$S2_{ij} := \frac{\sigma_{P^j}^2}{2}\mathrm{diag}H_{ij} \qquad (4.12)$$

to measure the **nonlinear influences** in each mesh node, analogous to the local sensitivity matrix. $\mathbf{S2}$ represents an approximation of the second term in the Taylor expansion in each mesh node reflecting the changes due to nonlinear (second order) effects.

An accumulated diagonal part of the Hessian matrix can be expressed as

$$\mathrm{diag}H^j := \frac{1}{\mathrm{Nnodes}} \sum_{i=1}^{\mathrm{Nnodes}} |\mathrm{diag}H_{ij}|, \quad j = 1, \ldots, \mathrm{Npar} \qquad (4.13)$$

and the **accumulated measure** $S2^j$ of parameter $P^j$ is given by

$$S2^j := \frac{\sigma_{P^j}^2}{2}\mathrm{diag}H^j. \qquad (4.14)$$

We define two **nonlinearity classes** $\mathrm{NLClass}_k, k = 1, 2$. $\mathrm{NLClass}_1$ contains all parameters whose influence on the criterion is supposed to be nonlinear. As opposed to that, NLClass$_2$ contains the parameters which influence the criterion only in a linear way. The parameter classification into nonlinearity classes proceeds as follows

$$\forall j \in \{1, \ldots, \text{Npar}\} :$$
$$J^j \leq c \, \text{diag} H^j \Rightarrow P^j \in \text{NLClass}_1, \qquad (4.15)$$
$$J^j > c \, \text{diag} H^j \Rightarrow P^j \in \text{NLClass}_2. \qquad (4.16)$$

Considering the Taylor expansion, an appropriate choice of $c$ is $c = \frac{\sigma_{Pj}}{2}$. If $J^j$ and $\text{diag} H^j$ are almost zero, the parameter $P^j$ will have no influence on the criterion at all, but this parameter will be assigned to NLClass$_1$. The importance classes reflect that this parameter is not influential, therefore, it will be neglected in further analysis. With $c = \frac{\sigma_{Pj}}{2}$ the equivalence

$$J^j \leq c \, \text{diag} H^j \Leftrightarrow S1^j - S2^j \leq 0 \qquad (4.17)$$

holds. This is referred to as **accumulated nonlinearity measure** in the following, since parameters $P^j$ fulfilling Equation (4.17) are assigned to the first nonlinearity class according to Equation (4.15).

In addition to the overall classification, local effects can be analyzed by pointwise comparing the entries of the Jacobian and the diagonal part of the Hessian matrix, i.e., if

$$|J_{ij}| < c \, |\text{diag} H_{ij}| \Leftrightarrow |S1_{ij}| - |S2_{ij}| < 0 \qquad (4.18)$$

with $c = \frac{\sigma_{Pj}}{2}$ holds, the dependence of the $j$-th parameter on the criterion in mesh node $N_i$ should be supposed to be nonlinear.

This new pointwise consideration of parameter sensitivities and nonlinearity measures can give valuable insight into process characteristics. This is specifically important in forming processes, where a lot of physical phenomena, e.g., intense thinning or folding, occur only locally. The local investigation enables the identification of the most critical areas of the forming mesh.

It is important to note that to assume linear behavior in the case of NLClass$_2$ and if Equation (4.18) is not fulfilled, will not always be true. It is possible that the Hessian matrix is not diagonally dominant. Especially, it can occur that **diag**$H = 0$, but the off-diagonal elements are large. Even if the entire Hessian matrix is almost zero, higher order derivatives could be large, so that the underlying function will be strongly nonlinear.

In order to ensure that in this case the parameters behave linear on the criterion, that is, at least the derivatives of second order are small, the full Hessian matrix has to be analyzed. This requires the computation of Npar(Npar$-1$)/2 additional entries of the Hessian matrix due to its symmetry. For this purpose, we have to add $4 \times$ Npar(Npar$-1$)/2 simulation runs to the star-point DoE to approximate the required second partial derivatives with second order FD schemes. The second partial derivative w.r.t. the $j$-th and $l$-th parameter in mesh node $N_i$ is defined as

$$
\begin{aligned}
H_{ijl} &:= \frac{\partial^2 g_i(\mathbf{P})}{\partial P^j \partial P^l} \\
&\approx \frac{g_i(\mathbf{P}^{j+1,l+1}) - g_i(\mathbf{P}^{j+1,l-1}) - g_i(\mathbf{P}^{j-1,l+1}) + g_i(\mathbf{P}^{j-1,l-1})}{4\sigma_{Pj}\sigma_{Pl}},
\end{aligned} \tag{4.19}
$$

where all other parameters except $P^j$ and $P^l$ are held fixed, see Equation (2.6). We use

$$
H_{jl} = \frac{\sigma_{Pj}\sigma_{Pl}}{\text{Nnodes}} \sum_{i=1}^{\text{Nnodes}} |H_{ijl}| \tag{4.20}
$$

as an accumulated approximation of the $j, l$-th entry of the Hessian matrix $\mathbf{H} = (H_{jl})_{j,l=1,\dots,\text{Npar}}$ scaled by the corresponding variation ranges. The vector of all accumulated linear sensitivities, given by Equation (4.3), scaled by the corresponding parameter variation range is given by $\widetilde{\mathbf{J}} = [\sigma_{P1}J^1, \dots, \sigma_{P\text{Npar}}J^{\text{Npar}}]^T$.

We define a **measure** $D$, similar to [20, 21], motivated by the Taylor expansion, as

$$
D := ||\widetilde{\mathbf{J}}||_2 - \frac{1}{2}||\mathbf{H}||_2 = ||\widetilde{\mathbf{J}}||_2 - \frac{1}{2}|\alpha|_{\text{max}}, \tag{4.21}
$$

where $|\alpha|_{\text{max}}$ denotes the maximal absolute eigenvalue of the Hessian matrix $\mathbf{H}$, and both, the linear and nonlinear influences, are already scaled by the corresponding variation range. Otherwise, the maximal absolute eigenvalue of the Hessian matrix has to be multiplied with $\frac{\sigma}{2}$, where $\sigma$ is the maximum of $\sigma_{Pj}, j = 1, \dots, \text{Npar}$, and all $\sigma_{Pj}$ should be of the same amount of magnitude [21].

The overall linear influence of all parameters is computed by

$$
||\widetilde{\mathbf{J}}||_2 := \sqrt{\sum_{j=1}^{\text{Npar}} (\sigma_{Pj}J^j)^2}. \tag{4.22}
$$

Note that the $L_2$-norm $|| \cdot ||_2$ and the induced matrix norm given by the spectral norm $|| \cdot ||_2$ is used in the definition (Equation (4.21)), cf. Section 2.1.

If $D$ is negative, the dependence of parameters on the criterion is supposed to be nonlinear, otherwise at least the second partial derivatives are small and the dependence on the criterion is assumed to be linear. We define a local measure, respectively, that writes

$$
D_i := \sqrt{\sum_{j=1}^{\text{Npar}} \sigma_{Pj}J_{ij}^2} - \frac{1}{2}|\alpha|_{i,\text{max}}, \tag{4.23}
$$

where $|\alpha|_{i,\max}$ denotes the maximal absolute eigenvalue of the Hessian matrix $\mathbf{H}_i$ scaled by the corresponding variation ranges in mesh node $N_i$.

| Linear importance classes | Nonlinearity classes |
|---|---|
| Based on $J_{ij}$, $J^j$ | Based on $\operatorname{diag} H_{ij}$, $\operatorname{diag} H^j$ *Optional:* if indicator of nonlinearity (Equation (4.15), Equation (4.18)) is false and additional simulation runs are practicable: based on $\mathbf{H}$, $D$, $D_i$ |

Table 4.1: Overview of parameter classification ingredients.

In summary, the parameter classification procedure computes the linear sensitivity measures and the indicators for nonlinearity, cf. Table 4.1. Only, if the inequalities Equation (4.15) and Equation (4.18) do not hold and, in addition, the simulation runtime is not a restrictive condition, the full Hessian matrix will be computed.

### 4.1.3 Total Importance Classes

We have analyzed the nature of effects due to parameter variations, that is, if the effects have a linear or nonlinear character, which results in the nonlinearity classes. In addition, we have assigned each parameter to a linear importance class, which reflects the amount of linear influence on the criterion.

It is possible that a parameter influence is nonlinear, but in total very small, so that this parameter variation can be neglected in further analysis. We summarize all these effects in order to prepare an efficient parameter space reduction. For that purpose, we define **total importance classes** TIClass$_k$, $k = 1, 2, \ldots$. The total importance classes are needed to order the parameters considered with respect to their overall influence on the criterion at least to second order derivatives (linear and nonlinear influences). Therefore, the (accumulated) first and second terms of the Taylor expansion, given by Equation (4.2), Equation (4.12), and Equation (4.4), Equation (4.14), respectively, are summed up. The **local total sensitivity** of parameter $P^j$ in mesh node $N_i$ is given by

$$\mathrm{TI}_{ij} := |\sigma_{P^j} J_{ij}| + |\frac{\sigma_{P^j}^2}{2} \operatorname{diag} H_{ij}| = |S1_{ij}| + |S2_{ij}|. \qquad (4.24)$$

We consider the **accumulated total sensitivity**

$$\mathrm{TI}^j := \sigma_{P^j} J^j + \frac{\sigma_{P^j}^2}{2} \operatorname{diag} H^j = S1^j + S2^j. \qquad (4.25)$$

The normalized total importance in $[0, 1]$ is given by

$$\widetilde{\text{TI}}^j := \frac{\text{TI}^j}{\sum_{j=1}^{\text{Npar}} \text{TI}^j}. \tag{4.26}$$

Then, the classification into total importance classes proceeds analogously to the classification into linear importance classes, described in Section 4.1.1. Therefore, the average proportion of the overall variation which can be explained by taking the variations of the parameters in the first $l$ total importance classes into account, $\text{TI}(l)$, can be computed by

$$\text{TI}(l) := \sum_{j,P^j \in \bigcup_{k=1}^{l} \text{TIClass}_k} \widetilde{\text{TI}}^j. \tag{4.27}$$

Analogously, the proportion of the overall variation which can be explained locally in mesh node $N_i$ by taking the variations of the parameters in the first $l$ total importance classes into account, $\text{TI}_i(l)$, can be computed by

$$\text{TI}_i(l) := \sum_{j,P^j \in \bigcup_{k=1}^{l} \text{TIClass}_k} \frac{\text{TI}_{ij}}{\sum_{j=1}^{\text{Npar}} \text{TI}_{ij}}. \tag{4.28}$$

## 4.1.4   Application of the Parameter Classification

The parameter classification procedure described is applied to two analytical test functions and the fundamental forming example introduced in Section 2.3. These examples illustrate the classification of parameters. Additionally, the benefit of the local measures is demonstrated.

### Analytical Test Functions

We look at the following analytical test functions to investigate the parameter classification procedure under various aspects:

$$f_1(x_1, \ldots, x_{\text{Npar}}) := 1.5 - 0.5\frac{1}{\text{Npar}} \sum_{i=1}^{\text{Npar}} \cos\left(2\pi(x_i - 0.5)\right), \tag{4.29}$$

$$f_2(x_1, x_2, x_3) := \sin x_1 + a \sin^2 x_2 + b x_3^4 \sin x_1. \tag{4.30}$$

The function $f_1$ is evaluated within $[-\pi/2, \pi/2]$, with $\text{Npar} = 2$, $\sigma_{P^j} = 0.1, j = 1, 2$ and $\text{Nexp} = 1000$. The accumulated (total) sensitivity measure shows equal influence of both parameters as expected by the function definition. Looking at the local nonlinearity measure, given by Equation (4.18), local effects can be detected correctly. Specifically, the nonlinearity of the function at the local minima and maxima is identified. This is demonstrated

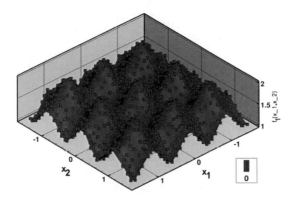

Figure 4.1: Local nonlinearity measure of parameter $P^1$. Blue color indicates nonlinear behavior.

in Figure 4.1, which shows the results of Equation (4.18) for the first parameter in each mesh node. The points shown are the sampling points. The influence of $P^1 = x_1$ is suppposed to be nonlinear, if $|S1_{i1}| - |S2_{i1}| < 0$ is fulfilled, which is indicated with blue color.

Second, we investigate the nonlinear Ishigami function $f_2$, presented in [51], which is a commonly used test function for benchmarking global sensitivity methods. Note that we sample the parameters $x_i, i = 1, \ldots, 3$ uniformly distributed within $[-\pi, \pi]$. Figure 4.2a represents a surface plot of the Ishigami function with $x_3 = \frac{\pi}{2}$. We evaluate $f_2$ using $a = 7$, $b = 0.1$, $\sigma_{Pj} = 0.1, j = 1, 2, 3$, and Nexp $= 1000$.

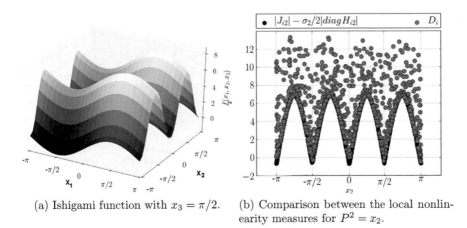

(a) Ishigami function with $x_3 = \pi/2$.

(b) Comparison between the local nonlinearity measures for $P^2 = x_2$.

Figure 4.2: Ishigami function and a comparison between both local nonlinearity measures.

The parameter classification into nonlinearity classes provides one non-linear parameter, i.e., it results in $\text{NLClass}_1 = \{x_2\}$. The Hessian matrix of the Ishigami function writes

$$\mathbf{H} = \begin{pmatrix} -\sin x_1(1 + 0.1x_3^4) & 0 & 0.4x_3^3\cos x_1 \\ 0 & -14\sin x_2 & 0 \\ 0.4x_3^3\cos x_1 & 0 & 1.2x_3^2\sin x_1 \end{pmatrix}. \quad (4.31)$$

The Hessian matrix is zero in $(0,0,0)$. Moreover, the Hessian matrix is diagonally dominant with $x_j \in [-\frac{3}{4}\pi, -\frac{\pi}{4}]$ and $x_j \in [\frac{\pi}{4}, \frac{3}{4}\pi]$, specifically,

$$\mathbf{H}(\frac{\pi}{2}, \frac{\pi}{2}, \frac{\pi}{2}) = \begin{pmatrix} -1.6 & 0 & 0 \\ 0 & -14 & 0 \\ 0 & 0 & 2.96 \end{pmatrix}. \quad (4.32)$$

For that reason, using the diagonal part of the Hessian matrix in the non-linearity measures as a weak approximation of the Hessian matrix works well, see Figure 4.2b. In particular, nonlinearities w.r.t. $x_2$ are detected around $x_2 \in \{-\pi, -\frac{\pi}{2}, 0, \frac{\pi}{2}, \pi\}$. The measure according to the diagonal part of the Hessian matrix, given by black points, builds a limit for the measure based on the full Hessian matrix. This would not be the case, if the matrix is not weakly diagonally dominant. In this example, both the local non-linearity measure based on the full Hessian matrix (gray) and the measure based on the diagonal part of the Hessian matrix (black) detect all local nonlinearities properly. To be more specific, the local nonlinearities at the extrema of the function w.r.t $x_1$ around $x_1 \in \{-\frac{\pi}{2}, \frac{\pi}{2}\}$ and w.r.t. $x_2$ around $x_2 \in \{-\frac{\pi}{2}, 0, \frac{\pi}{2}\}$ are detected properly, even if the local nonlinearity measure (Equation (4.18)) is used. That can be seen from Figure 4.3, which shows the evaluation of the local nonlinearity measure in each sampling point for parameter $P^1$ on the left and $P^2$ on the right hand side. Negative values (black) indicate nonlinear behavior. With respect to $P^3$ the local nonlinearity measure is about zero around $(0,0,0)$ and positive otherwise.

### Fundamental Forming Example

We come back to the forming example introduced in Section 2.3 and illustrate the parameter classification procedure developed. Recall that six material and process parameters are involved in this example. The parameter classification procedure reveals that all parameters varied influence both criteria, EPS and thickness, linearly almost everywhere. The influence on EPS measured by the accumulated sensitivity measure (Equation (4.4)) and the accumulated measure of nonlinearity (Equation (4.14)) is presented in Figure 4.4a. We explain in detail the parameter classification into linear importance classes by means of this forming example. The classification is illustrated in Figure 4.4b - 4.4d and proceeds as follows.

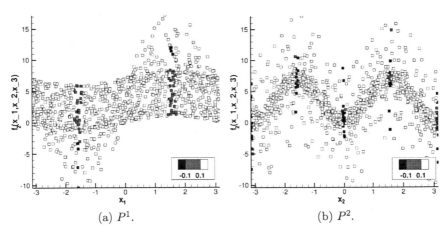

(a) $P^1$.                                    (b) $P^2$.

Figure 4.3: Local nonlinearity measure given by Equation (4.18) for the Ishigami function. Negative values (black) indicate nonlinear behavior.

The normalized accumulated sensitivity measures $\widetilde{S1}^j$ are computed for each parameter, $j = 1, \ldots, 6$. First, all parameters with a normalized accumulated sensitivity measure equal or greater than the average percentage of influence $(1/6)$ are assigned to the first linear importance class, i.e., $\text{LIClass}_1 = \{\mu\}$, see Figure 4.4b. Second, the normalization is recomputed with the remaining five parameters. According to Equation (4.7), a remaining parameter will be assigned to the second linear importance class (Figure 4.4c), if its influence is higher than $1/5$, i.e., $\text{LIClass}_2 = \{R_{90}, n\}$. Third, the procedure is iterated with the remaining three parameters (Figure 4.4d) resulting in $\text{LIClass}_3 = \{F^H, K\}$. Finally, the only remaining parameter $t$ is assigned to the last linear importance class $\text{LIClass}_4$. In total, the parameter classification procedure results in four linear importance classes.

The same procedure is applied to the accumulated total sensitivities (Equation (4.25)) resulting in the overall classification given in Table 4.2.

The friction $\mu$ and the strain hardening index $n$ influence the criterion EPS strongly. This is also the case considering the criterion thickness, but as expected, the parameter thickness itself has the strongest influence on the criterion thickness. Thus, the parameter classification into importance classes results in only slightly different assignments for the criteria EPS and thickness shown in Table 4.2.

Next, we compute the percentage of the overall variation which can be explained by taking only the first $l$ total importance classes into account, according to Equation (4.27). We obtain that already 71 percent (EPS) and 93 percent (thickness) of the overall variation can be explained by the

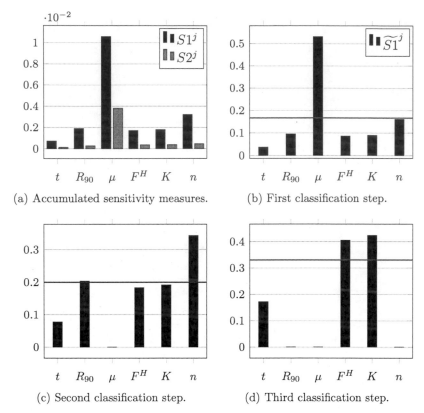

(a) Accumulated sensitivity measures.       (b) First classification step.

(c) Second classification step.             (d) Third classification step.

Figure 4.4: Forming example: Sensitivity measures and parameter classification into linear importance classes for the criterion EPS. The gray line corresponds to the average percentage of influence of the remaining parameters considered.

variations of the parameters assigned to the first and second total importance class, see Table 4.2. Considering the criterion thickness, 82 percent of the overall variation can be already explained by only taking the first total importance class into account. Therefore, the parameter space can be reduced without loss of much information. The procedure to reduce the parameter space is derived in Section 5.1.

Table 4.2 shows that no parameter is assigned to the first nonlinearity class, i.e., all parameters behave linear globally. Nevertheless, it is important to investigate the local influence of parameter variations on the criteria considered, because specific forming effects are often local in character. A comparison between the local linear sensitivity measure and the measure of nonlinear influences of parameter friction ($\mu$) on EPS is given in Figure 4.5. EPS is only locally and linearly influenced by $\mu$. The values of $S2_{i3}$

| Class | Parameter | EPS % | Parameter | Thickness % |
|---|---|---|---|---|
| LIClass$_1$ | $\mu$ | 0.53 | $\mu, t$ | 0.83 |
| LIClass$_2$ | $R_{90}, n$ | 0.79 | $K, n$ | 0.93 |
| LIClass$_3$ | $F^H, K$ | 0.96 | $F^H$ | 0.97 |
| LIClass$_4$ | $t$ | 1.0 | $R_{90}$ | 1.0 |
| NLClass$_1$ | — | | — | |
| NLClass$_2$ | $\mu, R_{90}, n, F^H, K, t$ | | $\mu, R_{90}, n, F^H, K, t$ | |
| TIClass$_1$ | $\mu$ | 0.57 | $\mu, t$ | 0.82 |
| TIClass$_2$ | $n$ | 0.71 | $K, n$ | 0.93 |
| TIClass$_3$ | $R_{90}, F^H, K$ | 0.97 | $F^H$ | 0.97 |
| TIClass$_4$ | $t$ | 1.0 | $R_{90}$ | 1.0 |

Table 4.2: Forming example: Parameter classification results for the criteria EPS and thickness. % denotes the average proportion of the overall variation explained by the union of the parameters up to the class considered.

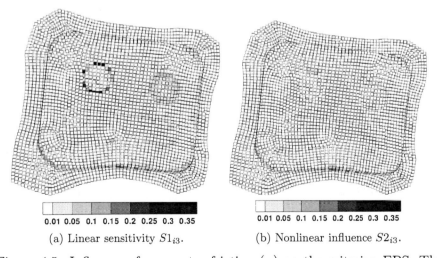

(a) Linear sensitivity $S1_{i3}$.

(b) Nonlinear influence $S2_{i3}$.

Figure 4.5: Influence of parameter friction ($\mu$) on the criterion EPS. The linear sensitivities are higher than the nonlinear influences in areas with significant overall influence.

belonging to the parameter $\mu$ are zero almost everywhere. The values of $S1_{i3}$ are significantly higher only at one of the secondary design elements, i.e., variations of $\mu$ lead only to local variations of EPS. Furthermore, the magnitude of linear sensitivities are much higher than the nonlinear influences in the region of the secondary design element with significant total influence. Thus, the behavior of the friction coefficient on EPS can be assumed to be linear in agreement with the classification in nonlinearity classes (cf. Table 4.2). Similar behavior of the parameter friction on the criterion thickness is obtained and, therefore, not shown here.

In order to validate the classification results based on the diagonal part of the Hessian matrix, we compute the measure $D$ based on the full Hessian matrix. This is usually not computed due to high simulation runtimes. That is, $4\mathrm{Npar}(\mathrm{Npar}-1)/2 = 60$ additional simulation runs have been performed. We obtain the measure $D = ||\widetilde{\mathbf{J}}||_2 - \frac{1}{2}|\alpha|_{\max} = 0.0115 - \frac{1}{2}0.0085 = 0.00728$, which is positive. Note that the measure is almost zero. This is expected, since the total influence is almost zero in the majority of mesh nodes. Thus, the linearity is confirmed considering influences up to second order. Recall that there could be influences due to higher order terms in the Taylor expansion which remain unknown. The computational effort to compute the measure $D$ is determined by the simulation runtime of the additional simulation runs required. Providing these are already performed the computation of the Hessian matrix, its maximal eigenvalue and the measure $D$ took only 0.07 seconds on a standard 3 GHz Linux computer.

## 4.2 Clustering Using the Nonlinearity Measure

The results of the parameter classification, specifically the results of the nonlinearity measure, can be used for clustering of the data set given. This results in disjoint subsets, so that points within a cluster are similar w.r.t. the resulting nonlinearity measure. That is, the parameter considered has similar influence on the criterion considered for all nodes of a specific cluster.

This domain decomposition is beneficial to subsequent analysis steps. Particularly, more advanced approximation methods, or a higher number of samples, need to be used only in the regions of interest, which reduces the overall computational effort.

The clustering of the local nonlinearity results proceeds as follows. The first parameter of NLClass$_1$ with the highest total influence, given by Equation (4.26), is considered. This parameter behaves nonlinear on the criterion considered and has a high overall influence. If all parameters belong to the second nonlinearity class, i.e., all parameters behave globally linear, the first parameter of TIClass$_1$ will be selected. The results of the local nonlinearity measure $|S1_{ij}| - |S2_{ij}|$ in each mesh node (Equation (4.18)) corresponding to the selected parameter $P^j$ represent the input data points for the clustering. Then, a common clustering algorithm is applied to this data set. We use the K-means clustering algorithm explained in Section 3.3.1. This algorithm finds clusters of the data points in corresponding mesh nodes with centers $c_{CL_k}, k = 1, \ldots, $ Nclust.

The number of clusters has to be determined a priori. We have observed that Nclust $= 3$ is suitable for most applications, because the points are separated therewith in clusters with almost no influence, linear influence, and nonlinear influence of the parameter considered. In specific cases, for example, the crash application presented in Section 7.3, the number of clusters may be increased slightly. Therewith, local, strongly nonlinear effects can be separated. In summary, the clusters represent disjoint domains of the geometry with respect to the nonlinearity of the underlying process.

This domain decomposition can be used to reduce the computational effort in subsequent steps of the PRO-CHAIN methodology. Especially, the propagation of variations, to be more specific, the interpolation of new designs and the computation of statistics, can benefit highly from this decomposition.

The strength of nonlinearity determines which interpolation method should be used to predict new designs. If the underlying process is linear, accurate results will be achieved with a linear interpolation method. If the influence of parameter variations is almost linear or moderately nonlinear, a metamodel with radial basis functions (RBFs) and moderate number of sampling points will be appropriate. If the influence of parameter variations is strongly nonlinear, much more sampling points will be required to generate a metamodel for the highly accurate prediction of the behavior of new designs. Therefore, on the one hand, it is possible to use different interpolation methods in different clusters. On the other hand, it is possible to use the same interpolation method, namely a RBF metamodel, with a hierarchical structure. Then, in the clusters corresponding to nonlinear results, local models with more samples are generated and combined with a global model which is less computationally expensive.

Moreover, it can be advantageous to evaluate the forecast model with different number of samples in different clusters for computing statistics. In clusters corresponding to a strong nonlinearity, more evaluations of the metamodel are required in order to achieve a good accuracy of the computed statistics. Whereas in clusters with no or only linear influence, much fewer evaluations are sufficient, which reduces the computational time needed to compute statistics.

Overall, different strategies in different clusters can particularly contribute to a reduction of the overall computational effort to predict new designs.

## Application of the Clustering to the Fundamental Forming Example

The parameter classification procedure assigns all parameters of the fundamental forming example to the second nonlinearity class. Hence, all parameters are assumed to be linear, as shown in Table 4.2. Therefore, we select the most important parameter, namely the friction coefficient $\mu \in \text{TIClass}_1$, and consider the criterion EPS. The K-means algorithm is applied to the results according to the local nonlinearity measure of $\mu$. As already stated, we have observed that three clusters often correspond to points with almost no influence, linear influence, and nonlinear influence of the parameter considered. Due to the linearity of all parameters, we expect good results with only two clusters. Hence, we compare the clustering with Nclust $= 2, 3$. When Nclust is set to two, a very small cluster around a secondary design element with 0.3 percent of the points is found, which corresponds to strong linear influence. The remaining cluster contains all other nodes with small or no influence. A better result with respect to the local classification result is obtained with three clusters. Figure 4.6 visualizes the mesh nodes

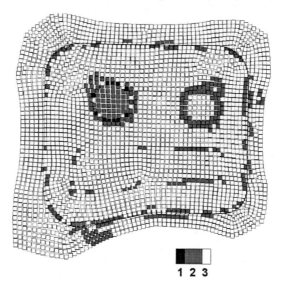

**1 2 3**

Figure 4.6: Clusters based on the nonlinearity measure of parameter friction with respect to the criterion EPS using Nclust = 3.

corresponding to the three clusters found. The first cluster contains only 0.3 percent of the data points, the second cluster contains 13.7 percent, and the third cluster contains 86 percent of the data points. The nodes in $\text{CL}_1$, with cluster center $c_{\text{CL}_1} = 0.28$, represent all nodes in which the parameter friction has strong linear influence. The nodes in $\text{CL}_2$, with cluster center $c_{\text{CL}_2} = 0.03$, contains all nodes in which the friction coefficient has little influence. Finally, $CL_3$ represents all nodes in which the variation of the friction coefficient has almost no influence. This is in agreement with the results of the linear and nonlinear influences shown in Figure 4.5, which are almost zero in the nodes of the third cluster.

## 4.3 Efficiency

We analyze the efficiency of the parameter classification procedure developed, in terms of the number of simulation runs, the computational complexity, and the memory required. The parameter classification is performed Ncrit times, that is, for each criterion separately. For simplicity, we set Ncrit = 1 in the following efficiency analysis.

### Number of Simulation Runs

The parameter classification procedure starts by generating a database containing simulation results. Since each simulation run is still very time consuming for the applications considered, the computational effort of the parameter classification procedure is determined by the simulation runtime. First of all, 2Npar + 1 simulations are performed to set up the database **M**. Then, the parameter classification based on entries of the Jacobian matrix and the diagonal part of the Hessian matrix can be performed using that database solely. Thus, the number of simulation runs is kept at a minimum. If the indicator for nonlinearity does not hold, only the full Hessian matrix will give a reliable sensitivity result (cf. Section 4.1.2). In order to determine the full Hessian matrix, Npar(Npar − 1)/2 additional entries due to its symmetry have to be computed. Furthermore, if we apply a finite difference scheme of second order, we will additionally need 4 simulation runs per entry. Even for a relatively small number of parameters, this results in a large amount of simulation runs. For example, Npar = 6 leads to 60 additional simulation runs. To this reason, the measure given by Equation (4.21) will only be feasible, if the number of parameters to investigate is very small and/or the simulation runtime is low.

After the parameter classification, the clustering is performed (cf. Section 4.2). It uses the computed local nonlinearity measure. Hence, no additional simulation runs are required.

### Computational Complexity

Considering the classification algorithm itself, we distinguish two cases analyzing the computational complexity. First, besides the local (total) sensitivity matrix, and the accumulated sensitivity measure, the nonlinear influences and accumulated measure based only on the diagonal part of the

Hessian matrix are computed. Second, additionally, the measure $D$ based on the full Hessian matrix is computed.

In the first case, the computations of the local sensitivity matrices and the corresponding accumulated measures, based on the diagonal part of the Hessian matrix solely, are of complexity $\mathcal{O}(\text{NparNnodes})$. The classification into linear and total importance classes, and nonlinearity classes is $\mathcal{O}(\text{Npar\#classes})$, where #classes denotes the maximum of the numbers of resulting linear and total importance classes. Note that usually #classes $\ll$ Npar holds. In the worst case, each class only contains one parameter, i.e., #classes = Npar. If the entries of the Jacobian and diagonal part of the Hessian matrix are pointwise compared according to Equation (4.18) in order to analyze local effects, the classification into nonlinearity classes results in $\mathcal{O}(\text{NparNnodes})$. Summarizing, the parameter classification procedure is mainly linear in the number of parameters and mesh nodes.

In the second case, we assume that the indicator of nonlinearity is false. Then, we additionally compute the measure $D$ based on the full Hessian matrix. The computation of the full Hessian matrix in each mesh node is $\mathcal{O}(\text{Npar}^2\text{Nnodes})$. Moreover, the maximal absolute eigenvalue of each Hessian matrix has to be computed. This results in an overall complexity of $\mathcal{O}(\text{Npar}^3\text{Nnodes})$ in the worst case. Recall that the measure based on the full Hessian matrix will only be computed, if the indicator of nonlinearity is false, and, many additional simulation runs are practicable.

The complexity of the clustering using the nonlinearity measure is dependent on the number of mesh nodes (Nnodes), the number of clusters (Nclust), and the number of iterations which have to be performed. For details of the complexity of the K-means algorithm used refer to Section 3.3.1.

## Memory Requirements

The parameter classification is based on a local sensitivity analysis, which needs to store a Nnodes $\times$ Nexp database $\mathbf{M}$, with Nexp = 2Npar + 1. However, since the sensitivity analysis is performed in each mesh node, the matrix is not stored entirely, but only one $1 \times$ Nexp vector is kept in memory. That is, the database can be processed on the fly. One row of the database containing values corresponding to a single mesh node is read into memory, analyzed, and the result is written. Therefore, the algorithm is able to deal efficiently with high dimensional databases. To store the resulting accumulated sensitivity measures together with the resulting classes $\mathcal{O}(\text{Npar})$ memory is required. Optionally, the local sensitivity matrices $\mathbf{S1}, \mathbf{S2}$, or the local nonlinearity measures $|S1_{ij}| - |S2_{ij}|$ can be stored with a memory complexity of $\mathcal{O}(\text{NnodesNpar})$.

The local nonlinearity measures are needed within the clustering procedure described in Section 4.2. For the clustering a Nnodes × 1 vector is loaded into memory corresponding to the local nonlinearity results of the parameter selected. As result, Nclust cluster centers are stored, together with a Nnodes × 1 vector containing the resulting cluster assignment in each mesh node.

## 4.4 Conclusions

We have developed a parameter classification procedure that assigns each parameter to a linear importance class, a nonlinearity class, and a total importance class. Furthermore, it computes linear and nonlinear influences locally on the entire mesh. Both types of classes, the importance classes and the nonlinearity classes, are a necessary basis to achieve the following two goals. First, a possible dimension reduction of the parameter space as well as an iterative extension of the database in case of nonlinear parameters is prepared. Second, the local consideration can identify interesting regions of the component. This gives a valuable insight into the local process behavior and supports identifying possible sources of failure. Therefore, with the approach developed both local and global impacts can be analyzed.

The major advantages of the approach proposed are the following. First, the assignment of parameters to linear and total importance classes is fully automated by the formal procedure described in Section 4.1.1. This procedure ensures that all parameters with a similar proportion of influence are assigned to the same importance class. The parameters of a single total importance class $TIClass_k$ should be either all retained in the analysis or all neglected. That is, the decision, if a parameter retains in the analysis, is based on its average importance relative to the other parameters. This is advantageous over a decision made by an arbitrary truncation after a certain amount of the overall influence which can be explained by the parameters selected.

Second, the approach proposed computes the distribution of the amount of linear and nonlinear influences due to parameter variations on the entire mesh. This gives an important hint to the local nature of parameter behavior. The subsequent clustering based on the computed local nonlinearity measure finds corresponding spacial regions of the component in which a parameter has similar influence on the criterion considered. Especially, local effects can be identified and separated. Hence, the subsequent prediction of the behavior of new designs and the computation of statistics can benefit from this decomposition by using more advanced approximation methods, or a higher number of samples, in the spacial regions of interest so that the accuracy of the forecast model is improved. If the number of simulations which can be performed is not restricted by the simulation runtime, the full Hessian matrix and the measure $D$ will be computed additionally. This approach ensures that only parameters with a small influence are neglected. The parameter classification approach newly developed differs considerably from state-of-the-art sensitivity studies, specifically performed in the automotive industry, for example, [83, 94, 39]. Particularly, the influences of parameter variations on criteria on fine meshes, i.e., large random fields, can be analyzed directly with the new approach. The importance to detect sensitivities on the entire mesh has been stated also in [1]. However, the authors suggest to use principal component analysis (PCA) (cf. Section 3.3.2) to compute the global correlations between parameters and the principal components of the model output. This incorporates all mesh nodes in the analysis. However, only linear influences can be detected and local effects cannot be revealed directly.

Third, the proportion of variation which can be explained by taking the union of the parameters of the first $k$ total importance classes into account is additionally estimated locally. Therewith, the parameter classification procedure gives an estimate of the expected local prediction quality, when a forecast model with a correspondingly reduced number of random parameters is used. If the forecast model, based on the automatic classification using the accumulated measures, is not expected to give suitable results in local areas of interest, the engineer can add parameters to the union of parameters taken into account, according to the local nonlinear influences computed.

In summary, the method proposed enables the usage of local estimates of the approximation quality, which is expected to provide better results in strongly nonlinear influenced regions than global ones. State-of-the-art measures of the approximation quality of metamodels are usually based on global estimates, for example, the coefficient of determination. Hence, they cannot resolve such local effects.

Finally, the approach proposed is very efficient and practicable also for problems based on computationally expensive simulation runs. The average computational time is linear in the number of mesh nodes and parameters considered. The number of required simulation runs is kept at a minimum. That is, the star-point DoE used for generating the database needs only 2Npar+1 samples. Based on this, all required measures for the classification can be computed. However, the parameter classification approach is limited to independent parameters, because interactions among parameters cannot be detected by varying only one parameter at a time. This limitation can be avoided by computing the full Hessian matrix. Interactions of higher order can only be computed with more computational effort.

Recall that global variance based sensitivity methods, for example, Sobol indices, are usually computed with Monte Carlo (MC) procedures. These require Nexp(Npar + 2) simulation runs [99] in order to compute all first order sensitivities. For example, for Npar = 15 and Nexp = 1000 already 17000 simulations are required. Hence, these methods will usually only be practicable in industrially relevant applications, if computed based on a metamodel (cf. Section 3.2.2). Another recently developed sensitivity method combines the search for a reduced parameter set with the search of the most appropriate approximation model, considering the coefficient of prognosis [82]. In this approach, metamodels for all possible subspaces have to be constructed and compared. Thus, this procedure results in a huge amount of computational costs and number of required simulation runs in order to generate many metamodels.

The local sensitivity analysis used in the approach proposed requires less computational effort than variance based sensitivity methods. Especially, the parameter classification approach proposed can be performed without the need of a metamodel. This is a major advantage over the sensitivity methods discussed above. Using the parameter classification procedure newly developed, an approximation model can be constructed using already the reduced set of parameters.

# Chapter 5

## Processing of the Database

The database contains an ensemble of simulation results, which builds the basis for subsequent analysis steps, especially, the construction of a forecast model. It is very important to **minimize the overall number of simulation runs**, since these determine the computational effort. Thus, the number of simulation runs are a main limiting factor within the analysis of process chains. Therefore, the database is created iteratively, based on the results of the classification procedure (cf. Chapter 4).

In this step, the parameter space can be reduced, which results in a **reduction of the database**. This is described in **Section 5.1**. The subset of the remaining important parameters which influence the criterion in a nonlinear way have to be analyzed in detail in order to model the nonlinearity properly by means of a metamodel. Thus, more simulation results may be needed. In this case, the database is extended iteratively. The ensemble of simulation results, which is stored in the created database, reflects the variations according to the important random parameters. This **iterative extension** of the database due to nonlinear parameters is presented in **Section 5.2**. Finally, the complete database is **compressed** by a basis transformation in order to reveal the main trends within the data. Therewith, the database is reduced further in order to minimize the remaining computational effort. We perform a lossy compression by means of a singular value decomposition (SVD) within the PRO-CHAIN methodology. This ensemble compression together with several error estimators are derived and investigated numerically in **Section 5.3**.

## 5.1   Parameter Space Dimension Reduction

The results of the parameter classification are used for the parameter space
dimension reduction. If the variation of a parameter does not influence the
criterion, at least to a certain amount, the variation of this parameter can be
neglected in further analysis. This means that this parameter is set constant
at its nominal value, and it is not a stochastic variable anymore. An error is
introduced in the model by neglecting parameter variations compared with
the full model containing all parameters. This error, $\varepsilon_\sigma \in [0,1]$, can be
expressed as a percentage of the overall resulting variation which cannot
be explained in the reduced model anymore. It depends on the goal of
the subsequent analysis and on the error bound which should be satisfied,
up to which amount of influence a parameter has to be considered in the
further analysis. Non influential parameters are assigned to the last total
importance class within the parameter classification procedure described in
Chapter 4. If the influence of a parameter is linear, computational cheaper
methods can be applied, for example, in the forecast model, compared with
strongly nonlinear parameters.

The parameter classification developed ensures that parameters with
a similar amount of influence are assigned to the same importance class.
Therefore, the parameter space dimension reduction considers all parameters of one total importance class uniformly. All parameters of that class
are either considered further or set to their nominal value.

The selection for setting parameters constant is controlled by a user
specified accuracy, $(1 - \varepsilon_\sigma) \in [0,1]$, which should be achieved. The relative error bound $\varepsilon_\sigma$ is usually chosen between 0.0 and 0.3, that is, at least
70 percent of the overall variation caused by parameter variations should
be taken into account on average.

We consider the total importance of each parameter and compute the
sum of the accumulated and normalized total sensitivities of the parameters
in the first importance class according to Equation (4.26). This results in
an average proportion of the overall variation that can be explained by the
parameters in the first total importance class. If this sum is equal or higher
than the required accuracy $(1 - \varepsilon_\sigma)$ all other parameters will be set to their
nominal value. If the threshold is not reached the parameters of the next
total importance class will be added and the comparison between the sum
of total sensitivities and the required accuracy is repeated iteratively until
the threshold is reached. This procedure to reduce the dimension of the
parameter space is presented in Algorithm 5.1.

The database $\mathbf{M}$ is reduced by the simulation results according to the
variations of the parameters fixed now. The columns of the database corresponding to variations of these parameters are deleted so that these variations are not considered in further analysis. After performing this pa-

---

**Algorithm 5.1** Procedure for parameter space dimension reduction

---

1: $k \leftarrow 1$
2: **while** $\sum_{j, P^j \in \bigcup_{i=1}^{k} \text{TIClass}_i} \widetilde{\text{TI}}^j < 1 - \varepsilon_\sigma$ **do**
3:    $k \leftarrow k + 1$
4: **end while**
5: $\forall j, P^j \notin \bigcup_{i=1}^{k} \text{TIClass}_i : P^j \leftarrow P^j_{\text{nom}}$

---

rameter space reduction, the database contains the simulation results of all remaining important random parameters and the result corresponding to the nominal parameter set. Therefore, the database has a reduced size by the amount of two times the number of fixed parameters compared with the original database. The database would have at most the same size as before, if all parameters remained influential.

The reduction procedure, given by Algorithm 5.1, is based on the variation explained on average. The parameter classification procedure provides additionally an estimate of the local variation explained, given by Equation (4.28). That is, if the expected local quality of a forecast model is not appropriate, the local estimate of the variation explained can be used in order to improve the expected forecast quality (cf. Section 4.4). For this purpose, the engineer can decide to add parameters to the set of parameters considered as random in the further analysis. These parameters should be selected based on the local distribution of the nonlinear influences. The database is adjusted accordingly as described above.

**Remark 5.1.** *In the following,* Npar *denotes the new number of random parameters consisting of the previous number of parameters reduced by the number of fixed parameters.* **M** *denotes the correspondingly reduced database.*

The remaining of this section illustrates the parameter dimension reduction procedure for the fundamental forming example introduced in Section 2.3. The reduction of the parameter space is based on the classification results shown in Section 4.1.4.

### Numerical Results for Fundamental Forming Example

We investigate two scenarios for each criterion considered based on the summed total sensitivities $\sum_{j, P^j \in \bigcup_{i=1}^{k} \text{TIClass}_i} \widetilde{\text{TI}}^j$ listed in Table 4.2.

In the first case, we analyze the criterion effective plastic strain (EPS) and aim at explaining on average at least 70 percent of the overall variation. In this case, it is sufficient to take all parameters in the union of $\text{TIClass}_1$ and $\text{TIClass}_2$ into account. That is, the random parameters $\mu$ and $n$ remain in the analysis, all simulation results belonging to other parameter variations are deleted from the database. This leads to a reduction of the parameter space dimension to Npar = 2, and a reduction of the database to one-third of its original size.

Second, if we want to obtain an average accuracy of 95 percent of the variation instead, the third total importance class has to be added. In this case, the resulting number of parameters remaining in the analysis is Npar = 5. This would lead only to a slight reduction of the parameter space, and, thus, of the database. The reduced database contains 11 out of the original 13 simulation runs resulting in a new size of the database of 85 percent of its original size.

Since this reduction procedure is based on the variation explained on average, we additionally investigate the local sum of total sensitivities. This leads to a local estimate of the expected forecast quality of a metamodel with correspondingly reduced number of parameters. We obtain that the proportion of variation explained is mostly already higher than the average expected 71 percent at the secondary design elements considering the sum of the first and second total importance class (Npar = 2) in Figure 5.1a. This is expected from the strong influence of the parameter friction ($\mu$) at one of the secondary design elements, cf. Figure 4.5. The variations at the borders of the geometry, that is, the domain of the pan which is mostly not highly deep drawn, cannot be explained well with solely the selected two parameters resulting in a locally lower expected accuracy than the average. Taking additionally the parameters of the third total importance class into

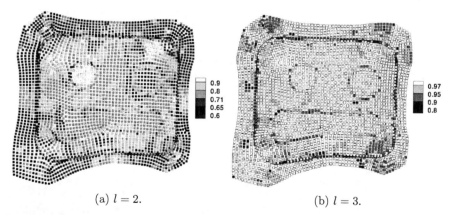

(a) $l = 2$.                          (b) $l = 3$.

Figure 5.1: Local proportion of the overall variation explained by the variations of the parameters in the first l total importance classes considering the criterion EPS.

account, the proportion of the overall variation explained strongly improves locally. In most points of the geometry more than 97 percent of the overall variation can be explained by taking these five parameters into account, as can be seen from Figure 5.1b.

In the second case, we analyze the criterion thickness and aim at explaining at least 70 percent of the overall variation on average. The parameters in the first total importance class can already explain 82 percent of the overall variation on average. Thus, the parameter space dimension is reduced to Npar = 2 and only the parameters $\mu$ and $t$ remain in the analysis.

If we want to obtain an accuracy of 90 percent of the variation on average, it is obvious from Table 4.2 that it will be sufficient to take all parameters in the union of TIClass$_1$ and TIClass$_2$ into account. This leads to a reduction of the parameter space to Npar = 4 parameters and a reduction of the database to two-third of its original size.

In summary, the parameter space can be reduced significantly without loss of much information in this examples, since few parameters account already for the majority of influences due to parameter variations. We come back to these estimators of the forecast quality, when evaluating the meta-models to predict the criteria considered in this example in Section 7.2.1.

# 5.2  Iterative Extension of the Database

In the previous section, the database has been reduced by setting non influential parameters to their nominal value. Therefore, the processed database now only represents the variations due to the remaining important and possibly nonlinear parameters. The important parameters which influence the criteria considered in a nonlinear way have to be characterized in detail in order to predict the criteria aimed at. Specifically, the three simulation runs for each nonlinear parameter, stored in the database so far, may be not sufficient in order to construct a forecast model with appropriate accuracy. Hence, more simulation runs may be required to extend the database iteratively until the forecast quality is high enough. Therefore, it is very important that the design of experiments (DoE) method used enables an iterative extension of the set of sampling points, like the one used in this work, cf. Section 3.1.

The iterative extension of the database distinguishes the following cases based on the parameter classification into nonlinearity classes described in Section 4.1.2:

1. All remaining important parameters are assigned to the second nonlinearity class, that is, they behave linear on average.

2. All remaining important parameters are assigned to the first nonlinearity class, that is, they behave nonlinear on average.

3. Both nonlinearity classes contain some of the remaining important parameters.

In the first case, a linear interpolation is expected to give already appropriate results. Therefore, the initially processed database is sufficient, and a further extension is not needed.

In the second case, the DoE has to be extended in each direction of the parameter space. Since, the radial basis function (RBF) metamodels used as forecast models suppose uniformly distributed sampling points, a random DoE is applied to extend the database. To be more specific, the additional sampling points are selected randomly and uniformly distributed within the complete remaining parameter space. To ensure a uniform distribution, including the already existing samples in the database, for each new randomly selected sampling point, the minimal distance to the already existing or selected points is computed. If this distance is below a certain threshold, the new point will be omitted. This procedure is repeated until the total number of sampling points required is reached.

In the final case, the same procedure as in the second case can be applied to ensure uniformly distributed sampling points. However, an appropriate forecast quality might be already achieved with fewer samples, if only few parameters behave nonlinear. In this case, hierarchical metamodels can be generated, which use more sampling points in the nonlinear directions of the parameter space. That is, the procedure of the second case is applied to subsets of the parameter space including only nonlinear parameters. An extension to adaptive hierarchical metamodels has been developed in [14] for the RBF metamodels used in this work.

The total number of sampling points (Nexp), and, thus, the total number of simulation runs required, in the second and third case is determined by Equations (3.22) and (3.23). In order to minimize the number of simulation runs required, the constant $C$ is also minimized. Hence, we use $C = 2$ and at least Nexp = 20 within the first iteration.

Furthermore, an iterative refinement of the DoE in regions in which the model tolerance, given in Equation (3.21) (cf. Section 3.4), is high can improve the forecasting quality locally. This iterative improvement is also useful in adaptive hierarchical metamodels. For refinement procedures in the latter case refer to [14] and references therein.

The extended database is processed further by an ensemble compression described in the following section. Then, the PRO-CHAIN methodology continues with the setup of the forecast models, derived in Chapter 6. If the prediction accuracy of the forecast models created is insufficient, the processing of the database, specifically, the extension and the ensemble compression, will be iterated in order to improve the forecasting quality.

In this case, the constant $C$ is increased. Details of this procedure can be found in Section 6.3.

If other types of approximation methods are used, which are not based on RBFs, the total number of sampling points has to be specified according to the assumptions of the method used.

## 5.3    Ensemble Compression of the Database

The previous modifications of the database either reduce the database by removing results belonging to non-influential parameters or extend the database by adding results belonging to strongly nonlinear behaving parameters. The so constructed database **M** contains an ensemble of Nexp simulation results according to a certain (random) DoE. This ensemble of simulation results reflects the variations of the important Npar random parameters.

**Remark 5.2.** *Note that* Npar *denotes the new number of random parameters consisting of the previous number of parameters reduced by the number of fixed parameters, cf. Section 5.1.* Nexp *denotes the new number of simulation runs which is (a part of) the previous* 2Npar + 1 *simulations extended by the number of additional simulations due to nonlinearities, cf. Section 5.2.* **M** *denotes the correspondingly processed database of size* Nnodes × Nexp.

The aim of the ensemble compression is to reduce the database further. This step is important in order to minimize the remaining computational effort for a subsequent analysis. Especially, it is important for the generation of a forecast model which can be evaluated fast.

We use a compression of the entire database in a physical way, that is, the principal trends of the database should be revealed. This is achieved by a basis transformation that maintains the important information while reducing the amount of memory required. For this purpose, we use a SVD of the database, computed with the fast algorithm according to Equation (3.10). This SVD is then also used for the acceleration of the RBF metamodel predicting new designs, which is introduced in Section 6.1.

In order to apply a SVD to the database **M**, the database should be normalized by subtracting the mean of the experiments of each row. The normalized database is denoted by $\widehat{\mathbf{M}} = (\widehat{M}_{ij})_{i=1,\ldots,\text{Nnodes},j=1,\ldots,\text{Nexp}}$, i.e., for each entry

$$\widehat{M}_{ij} := M_{ij} - \frac{1}{\text{Nexp}} \sum_{j=1}^{\text{Nexp}} M_{ij}. \tag{5.1}$$

A direct implication of this normalization is the following.

**Corollary 5.3.** *Let* $\widehat{\mathbf{M}} = \mathbf{U}\boldsymbol{\Lambda}\mathbf{V}^T$ *be the SVD of the database* $\mathbf{M}$ *with subtracted mean of the experiments. Then, it holds*

$$\lambda_{\text{Nexp}} = 0 \qquad and \tag{5.2}$$

$$\mathbf{V}_{\text{Nexp}} = (\frac{1}{\sqrt{\text{Nexp}}}, \dots, \frac{1}{\sqrt{\text{Nexp}}})^T, \tag{5.3}$$

*with the right singular vector* $\mathbf{V}_{\text{Nexp}}$ *eigenvector of* $\widehat{\mathbf{M}}^T\widehat{\mathbf{M}}$ *associated with* $\lambda_{\text{Nexp}}$.

Let $\widehat{\mathbf{M}}^k$ be the rank-$k$ approximation of the normalized database, given by Equation (3.11), that is, all singular values with $l = k + 1, \dots, \text{Nexp}$ are set to zero. The squared error by truncating the SVD after the $k$-th singular value can be expressed in terms of the Frobenius norm, according to Equation (3.12), as the sum of squares of the omitted singular values.

**Remark 5.4.** *The SVD can be applied to the database* $\widetilde{\mathbf{M}}$, *containing results and corresponding geometry over time, as well, provided that the different criteria and coordinates are of similar size. This can be achieved, for example, by a scaling to their root mean square (RMS) which is the square root of the average of the squares of the entries.*

We use the SVD not only for ensemble compression but also to accelerate the prediction of the behavior of new designs in Section 6.1.1. In this case, the SVD is evaluated for each new design. Specifically, thousands of evaluations are performed estimating the statistics of the model. For this reason, we are interested in selecting $k$ in the rank-$k$ approximation as small as possible, so that the matrix multiplications are much faster compared with taking all singular values into account. Therefore, we focus at minimizing the average approximation error due to the SVD performed. In the following, we investigate possible estimators of this average approximation error in detail.

The root mean squared error (RMSE) is a frequently used measure of the average error between a prediction, that is, an estimator $\widehat{y}$, and the actually observed value $y$. It is defined for Nnodes predictions as

$$\text{RMSE} := \sqrt{\frac{1}{\text{Nnodes}} \sum_{i=1}^{\text{Nnodes}} (\widehat{y}_i - y_i)^2}. \tag{5.4}$$

Let an arbitrary column of the normalized database, representing an arbitrary experiment, be denoted by $\widehat{\mathbf{M}}_p$, which is a vector of size Nnodes. Let the deviation due to the rank-$k$ approximation in this arbitrary experiment be defined by $\widehat{d\mathbf{M}}_p$. Then, the $i$-th entry of the deviation is given by

$$\left(\widehat{d\mathbf{M}}_p\right)_i = \left(\widehat{\mathbf{M}}_p^k - \widehat{\mathbf{M}}_p\right)_i = \sum_{l=k+1}^{\text{Nexp}} U_{il}\lambda_l V_{pl}. \tag{5.5}$$

Since the columns of $\mathbf{U}$ are orthonormal, it follows

$$\sum_{i=1}^{\text{Nnodes}} \left(\widehat{d\mathbf{M}}_\mathbf{p}\right)_i^2 = \sum_{l=k+1}^{\text{Nexp}} \lambda_l^2\, V_{pl}^2. \tag{5.6}$$

With these results, the RMSE of the arbitrary experiment $\widehat{d\mathbf{M}}_p$ can be expressed as

$$\text{err}_{\text{RMSE}}(k) = \sqrt{\frac{1}{\text{Nnodes}} \sum_{i-1}^{\text{Nnodes}} \left(\widehat{d\mathbf{M}}_p\right)_i^2}$$

$$\underset{(5.6)}{=} \sqrt{\frac{1}{\text{Nnodes}} \sum_{l=k+1}^{\text{Nexp}} \lambda_l^2\, V_{pl}^2}. \tag{5.7}$$

This shows that the RMSE of an arbitrary experiment is dependent on the matrix $\mathbf{V}$. Especially, the matrix $\mathbf{U} \in \mathbb{R}^{\text{Ndata}\times\text{Nexp}}$ has not to be computed in order to estimate the average approximation error. Thus, we estimate the entries of $\mathbf{V}$ in order to derive a general estimator of the average approximation error introduced by truncating the SVD after $k$ singular values.

An important phenomenon in high dimensional spaces is the concentration of measure [30, 71], which is one of the phenomena associated with the curse or blessings of dimensionality, when dealing with high dimensional data. One result is that the volume $\mathcal{V}$ of a Nexp-dimensional ball of fixed radius $r$ approaches zero as Nexp tends to infinity, since the volume is expressed as

$$\mathcal{V}_{\text{Nexp}}(r) = \frac{\pi^{\frac{\text{Nexp}}{2}} r^{\text{Nexp}}}{\Gamma\left(1 + \frac{\text{Nexp}}{2}\right)}, \tag{5.8}$$

where $\Gamma(\cdot)$ denotes the Gamma function. Furthermore, the volume of the ball gets concentrated near the equator for Nexp large. This means that a typical coordinate of a random point in a Nexp-dimensional unit ball is of size $1/\sqrt{\text{Nexp}} \ll 1$ [71].

Motivated by these observations, we restrict the range of $\mathbf{V}$ in order to derive appropriate error estimators from $\text{err}_{\text{RMSE}}(k)$. Assume the domain of $\mathbf{V}$ is restricted to one of the following domains

$$\forall p, l = 1, \ldots, \text{Nexp} :$$

$$|V_{pl}| \leq 1, \tag{5.9}$$

$$|V_{pl}| \leq \frac{1}{\sqrt{\text{Nexp}}}. \tag{5.10}$$

The first constraint is always fulfilled, because the matrix $\mathbf{V}$ is orthogonal by definition. The second constraint represents the maximal hypercube which is inscribed in the Nexp-dimensional unit sphere. In this case, we only consider points near the equator assuming that this holds for the majority of all entries of $\mathbf{V}$. In particular, the average of all absolute entries of $\mathbf{V}$ is expected to be bounded by $\frac{1}{\text{Nexp}}$ for Nexp sufficiently large, due to the above described phenomenon of concentration of measure.

Maximizing over the experiments, we obtain with Corollary 5.3, the error formula given by Equation (5.7), and for all $l$, $V_{pl} = 1$ and $V_{pl} = 1/\sqrt{\text{Nexp}}$, respectively, the error estimators

$$\text{err}_1(k) = \sqrt{\frac{1}{\text{Nnodes}} \sum_{l=k+1}^{\text{Nexp}} \lambda_l^2}, \tag{5.11}$$

$$\text{err}_2(k) = \sqrt{\frac{1}{\text{NnodesNexp}} \sum_{l=k+1}^{\text{Nexp}} \lambda_l^2}. \tag{5.12}$$

Note $\text{err}_2(k)$ coincides with the average error over all experiments using the orthogonality of $\mathbf{V}$. $\text{err}_1(k)$ represents the average error over all points of a single, arbitrary experiment. In summary, both derived error estimators predict average absolute error values.

Obviously, it holds $\text{err}_2(k) \leq \text{err}_1(k)$. In this work, we consider databases with Nexp $\geq 20$, cf. Section 5.2. Hence, due to the concentration of measure described, the majority of absolute entries of $\mathbf{V}$ is smaller than 1. Thus, the estimator $\text{err}_1(k)$ usually strongly overestimates the average approximation error due to the rank-$k$ approximation of the database. Instead, the estimator $\text{err}_2(k)$ is a good estimator of the average approximation error.

The SVD can be used as lossy compression of the database. That is, the truncation of the SVD leads to an ensemble compression of the database, and at the same time to a loss of information. This loss of information is controllable with the above error estimators. It depends on how many singular values are omitted. Therefore, it is essential to find a rule to specify $k$ suitable to achieve a certain accuracy. That is, we need to define $k$ so that a given error threshold $\varepsilon_{\text{SVD,abs}}$ is not exceeded, i.e.,

$$k : \quad \text{err}_j(k) < \varepsilon_{\text{SVD,abs}}, \tag{5.13}$$

with a fixed $j \in \{1, 2\}$. This threshold represents the averaged amount of loss of information which is acceptable. As already stated, $\mathrm{err}_1(k)$ refers to the average error of a single experiment, and $\mathrm{err}_2(k)$ to the average error of all experiments in the database. Therefore, the error threshold depends on the error estimator used and should be specified dependent on the application considered. For example, an appropriate error threshold to be used with $\mathrm{err}_2(k)$ can be, for example, five percent of the average variation present in the matrix.

**Remark 5.5.** *We use the derived estimator of the average approximation error due to a truncation of the SVD in the estimation of the overall average prediction error for a single processing step, in Section 6.4.*

To address the drawbacks of commonly used rules to choose the number of singular values retained (cf. Section 3.3.2), we have derived a stopping rule, given by Equation (5.13), based on the above error estimators and a user specified error threshold. This stopping rule is used directly within the PRO-CHAIN methodology as described in Sections 6.1.1 and 6.2.

Moreover, the truncation of the singular values leads to savings in memory, because only the decomposition with the relevant columns of $\mathbf{U}$ needs to be saved compared with the entire decomposition.

**Definition 5.6** (Data compression ratio). *The **data compression ratio** $C_R$ is defined as*

$$C_R = \frac{s_{unc}}{s_c}, \tag{5.14}$$

*where $s_{unc}$ denotes the uncompressed size, and $s_c$ denotes the compressed size of the data, respectively. That is, it quantifies the reduction in data representation size.*

For example, a data compression ratio of three means that the compressed database requires only one-third of its original size. By truncating the SVD after $k$ singular values the data compression ratio is given by

$$C_R = \frac{\mathrm{Nexp}}{k + 1}, \tag{5.15}$$

because only the first $k$ columns of $\mathbf{U}$ need to be computed and stored in combination with the previously subtracted row average, and the matrices $\mathbf{\Lambda}$ and $\mathbf{V}^T$, rather than the full SVD. Additionally, the time required for computing the SVD itself is reduced by computing only the first $k$ columns of $\mathbf{U}$. Moreover, each evaluation of the metamodel can be accelerated by using the truncated SVD, as shown in Section 6.1.1. This is beneficial, when the computational time of the SVD is negligible compared with the time of numerous evaluations, cf. Section 6.5.

**Numerical Comparison of Error Estimators**

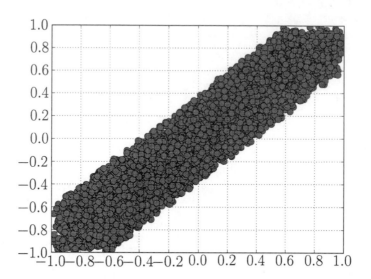

Figure 5.2: Specific structure of the test matrix with Nexp = 50: Entries of column 3 versus column 45 (chosen arbitrarily).

We compare the derived error estimators (Equations (5.11) - (5.12)) numerically on the basis of matrices with a specific structure similar to the structure of the considered databases. These Nnodes × Nexp matrices are generated as follows. We consider 4 test matrices with Nnodes = 10000 each, and with different Nexp ∈ {20, 50, 100, 200}. The smallest number of columns is determined as Nexp = 20, since this is the minimal number of experiments required to setup the metamodel used in this work, cf. Section 5.2. The first column of the matrix $\mathbf{A}$ is filled with random numbers uniformly distributed within $[-1, 1]$. Therefore, the first column can be identified with the results of a nominal experiment. The other columns should be identified with results due to parameter variations. Thus, the remaining Nexp − 1 columns of the matrix are generated with entries $A_{ij} = A_{i1} + \beta_i$, where $\beta_i$ is a random number uniformly distributed within $[-0.2, 0.2]$. The specific structure is illustrated in Figure 5.2. For each matrix, we extract the row mean, and compute the SVD of $\mathbf{A}$ and both estimators of its average error derived.

We start with investigating the entries of $\mathbf{V}$ in order to analyze the concentration of measure phenomenon described above. The Figure 5.3 shows exemplary the projection vectors $\mathbf{V}_1$ versus $\mathbf{V}_2$ belonging to the largest singular values of the SVD of $\mathbf{A}$ with different dimensions Nexp. The other projection vectors show a similar behavior. It holds $|V_{pl}| < 1, p, l = 1, \ldots, \text{Nexp}$,

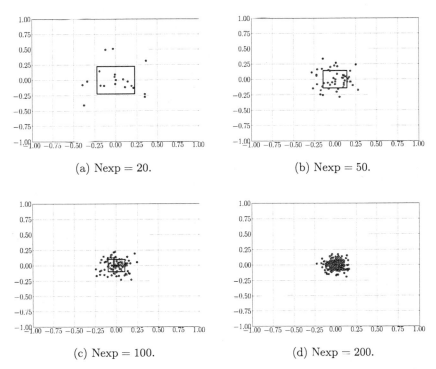

Figure 5.3: Projection vectors $\mathbf{V}_1$ versus $\mathbf{V}_2$ of the SVD of the different matrices considered. The rectangle shown restricts the domain $|V_{pl}| < 1/\sqrt{\text{Nexp}}$.

since $\mathbf{V}$ is orthogonal by definition. Furthermore, the figure illustrates that a much more compact region than the unit sphere is occupied by the projections of the experiments in high dimension. The rectangle shown in the figure restricts the domain $|V_{pl}| < 1/\sqrt{\text{Nexp}}$. In particular, this shows that the average of the entries $|V_{pl}|$ fulfill this condition. Moreover, this compact region gets smaller with increasing dimension. Thus, the effect of mass concentration in high dimensional spaces described above is observed also for the $\mathbf{V}$ matrices considered.

This shows that, on the one hand, the derived estimator $\text{err}_2(k)$ is a good predictor of the average error in this case, since the Equation (5.10) is fulfilled by the majority of the entries. On the other hand, the estimator $\text{err}_1(k)$ usually overestimates the average error strongly, since the Equation (5.9) is a much too rough condition for the majority of the entries. After dimension reduction due to the SVD, the entries $V_{pl}$ should belong to the compact region described. In this case, the estimator $\text{err}_2(k)$ gives a sharp estimator of the average error.

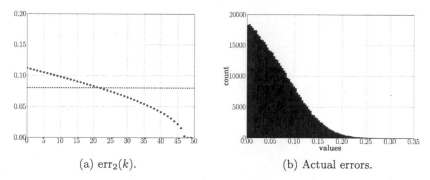

(a) $\mathrm{err}_2(k)$.                    (b) Actual errors.

Figure 5.4:  Error estimator $\mathrm{err}_2(k)$ for the test matrix with Nexp $= 50$ together with a histogram of the actual errors using $k = 23$ singular values. The dotted black line corresponds to the error threshold $\varepsilon_{\mathrm{SVD,abs}} = 0.08$.

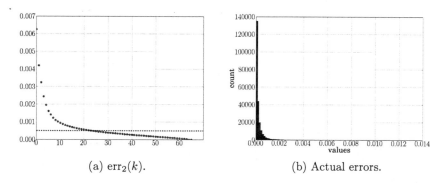

(a) $\mathrm{err}_2(k)$.                    (b) Actual errors.

Figure 5.5:  Error estimator $\mathrm{err}_2(k)$ for the fundamental forming example together with a histogram of the actual errors using $k = 24$ singular values and Nexp $= 66$. The dotted black line corresponds to the error threshold $\varepsilon_{\mathrm{SVD,abs}} = 0.0005$.

This is demonstrated exemplary for the matrix $\mathbf{A}$ with Nexp $= 50$ experiments. In detail, we accept an absolute average error of $\varepsilon_{\mathrm{SVD,abs}} = 0.08$, which corresponds to 20 percent of the variation in each row of the matrix. In this case, the estimator $\mathrm{err}_2(k)$ of the average error indicates to retain $k = 23$ singular values, as illustrated in Figure 5.4. The estimator $\mathrm{err}_1(k)$ is much higher, it results in $\mathrm{err}_1(23) = 0.56$. To assess the quality of these estimators, we compute the actual error. That is, the difference between $\mathbf{A}$

and $\mathbf{A}^{k=23}$ is computed in each entry. The histogram of these actual errors is illustrated in Figure 5.5b. The average is given by 0.054, which is smaller than $\mathrm{err}_2(23) = 0.079$. This demonstrates that the estimator $\mathrm{err}_2(k)$ provides a very good upper bound of the actual average error, whereas the other estimator strongly overestimates the average error. Indeed, $\mathrm{err}_1(23) = 0.56$ is higher than the actual maximal error.

This behavior is also confirmed by the applications considered. We exemplify this by the fundamental forming example introduced in Section 2.3. Figure 5.5 compares the estimator $\mathrm{err}_2(k)$ with the actual approximation error using $k = 24$ singular values. The histogram of the actual differences between the database and its rank-24 approximation shows an actual average error of 0.0001. It holds $0.0001 < \mathrm{err}_2(24) = 0.00049 < \mathrm{err}_1(24) = 0.004$. In summary, the estimator $\mathrm{err}_2(k)$ gives appropriate prediction results of the average error for the high dimensional applications considered. Thus, we use this estimator to determine the number of singular values retained, when accelerating the forecast model in Section 6.1.1.

# Chapter 6

---

# Forecast Model and Propagation of Variations

---

The database processed iteratively (cf. Chapter 5) builds the basis for the construction of appropriate forecast models. The prediction of out-of-sample points, called new designs, by means of **fast and accurate forecast models** enables the approximation of the probability distribution function. Thus, **statistical information of the processing step** is provided without additional time-consuming simulation runs. This statistical information can be used in a subsequent robust optimization.

We introduce the acceleration of state-of-the-art radial basis function (RBF) metamodels used to approximate new designs in **Section 6.1**. This acceleration is essential in order to deal with high dimensional simulation results efficiently. With this method, the possibly deformed geometry and important resulting criteria corresponding to a change in the parameter set are provided directly on the entire mesh, without additional simulation runs. Furthermore, possible failures of the component arising in a crash processing step are commonly not locally predictable with state-of-the-art metamodels. However, the **forecast of possible failure initiation** is absolutely essential to assess the results of a crash processing step. Therefore, we have derived a procedure to deal with deleted mesh elements representing this failure, resulting in a generalization of the accelerated RBF metamodels to crash processes.

In order to achieve a robust design, it is important to take all effects due to parameter variations into account. Especially, statistical information should be included in the set of optimization criteria. If the underlying process is nonlinear, appropriate robustness measures, for example, based on local quantiles, will be preferable to standard global measures as the

95

mean and standard deviation accumulated on the mesh. Therefore, the accelerated metamodel is evaluated to approximate the probability distribution function of the criteria considered by means of the computation of quantiles. An efficient way to compute statistics of the metamodel is derived in **Section 6.2**, so that it becomes possible to estimate these locally in each mesh node.

Bringing the previous PRO-CHAIN components together, namely the parameter classification procedure, the iterative processing of the database and the fast, high-quality forecast models, it is now possible to efficiently propagate all relevant scatter information from one processing step to the next. That is, an ensemble of vector-valued results representing the local, important variations in the result of the current processing step can be mapped to the next processing step. This allows taking both, the **history of the process and all relevant variations of the history**, into account in subsequent processing steps. The strategy of mapping the relevant information and their usage in subsequent processing steps is outlined in **Section 6.3**.

As described above, the prediction of the behavior of new designs and the computation of statistics is the final step within the analysis of a single processing step. To generate an appropriate database for these forecast models, several dimension reduction and approximation steps have been performed. Thus, several sources of possible errors have to be taken into account in order to assess the accuracy of the final prediction result. This **quality control** is subject of **Section 6.4**. Especially, the prediction error within a single processing step is derived theoretically and validated numerically. In addition, an estimator of the prediction error of the entire process chain is derived.

In the following, the efficiency of the forecast models developed and the algorithm to compute statistics is discussed in **Section 6.5**.

This chapter concludes with a discussion of the advantages and limitations of the forecast model proposed in **Section 6.6**.


## 6.1   Approximation of New Designs

Out-of-sample-points, that is, new designs representing realizations of the criteria with parameter sets which have not been simulated so far, can be predicted with the PRO-CHAIN methodology. Therefore, RBF metamodels are used as forecast models. To enable a statistical evaluation of the forecast model locally in an appropriate time, an acceleration of the RBF metamodel by the ensemble compressed (and propagated) database is introduced in Section 6.1.1.

In order to allow a meaningful assessment of the crash processing step, we have developed these metamodels further so that a local prediction of failure initiation becomes possible. The approach developed to deal with deleted mesh elements is presented in Section 6.1.2.

## 6.1.1 Radial Basis Function Metamodel Accelerated by a Singular Value Decomposition

The forecast model used to predict the criteria considered is a radial basis function metamodel according to Equation (3.18). We use the multiquadric function and polynomial detrending as described in Section 3.4.

Recall that an arbitrary processing step of the process chain, described on the entire mesh, is defined by the function vector $\mathbf{g} := [g_1, \ldots, g_{\text{Nnodes}}]^T$. The available simulation results $\mathbf{g}(\mathbf{P}_l)$ corresponding to the sampling points, i.e., parameter sets, $\mathbf{P}_l, l = 1, \ldots, \text{Nexp}$, are determined by the iterative generation procedure described in Chapter 5. Then, these results are stored in the processed database $\widetilde{\mathbf{M}}$. For simplicity, we set Nts = 1, Ncrit = 1 and dim = 3. In this case, 4Nnodes is the number of rows of the database.

In a first phase, only the weights $\mathbf{w}(\mathbf{P})$ of the metamodel, given by Equation (3.17), are computed. To accelerate the metamodel, the processed

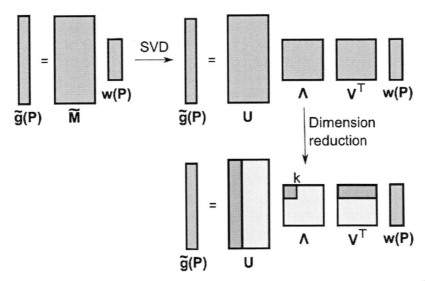

Figure 6.1: RBF metamodel accelerated with the SVD of the database $\widetilde{\mathbf{M}}$. Light gray areas represent entries which are set to zero within the dimension reduction procedure of the SVD.

database $\widetilde{\mathbf{M}}$ is replaced by its rank-$k$ approximation $\widetilde{\mathbf{M}}^k$, as suggested in [88,

20]. The rank-$k$ approximation results from the ensemble compression of the database by means of a singular value decomposition (SVD) performed in Section 5.3, i.e., all singular values with $l = k + 1, \ldots, \text{Nexp}$ are set to zero.

In the next phase, the results of the rank-$k$ approximation of the database, instead of the sampling results themselves, are combined with the metamodel weights in order to approximate the criterion for an arbitrary given parameter set within the parameter space. This procedure corresponds to a metamodeling of $\mathbf{V}^T$ and a projection of the resulting approximation in the space of the principal trends spanned by $\mathbf{U}\Lambda$, as illustrated in Figure 6.1.

The accelerated metamodel can be written as

$$\forall i = 1, \ldots, \text{Nnodes}:$$

$$\widetilde{g}_i(\mathbf{P}) \underset{(3.16)}{=} \sum_{l=1}^{\text{Nexp}} g_i(\mathbf{P}_l) w_l(\mathbf{P}) + \Psi_{\text{Dd}(\mathbf{P})} \tag{6.1}$$

$$\underset{SVD}{\approx} \sum_{l=1}^{k} U_{il} \lambda_l \widetilde{V}_l, \ \widetilde{V}_l = \sum_{j=1}^{\text{Nexp}} V_{jl}^T w_j(\mathbf{P}) + \Psi_{\text{Dd}(\mathbf{P})},$$

where $\Psi_{\text{Dd}}(\mathbf{P})$ specifies the polynomial detrending (cf. Equation (3.18)). The forecast model for the entire processing step $\mathbf{g}$ is determined by $\widetilde{\mathbf{g}} = [\widetilde{g}_1, \ldots, \widetilde{g}_{\text{Nnodes}}]^T$, respectively. Note that the metamodeling of $\mathbf{V}^T$ is independent of the mesh nodes. Due to the truncation of the SVD after the $k$-th singular value an approximation error is inserted, which can be controlled by choosing $k$. Several estimators of this error have been discussed in Section 5.3.

**Remark 6.1.** *If the coordinates are not needed, only the criteria will be approximated in order to save memory and computational time. In this case, using $Y_l^i$ and $\mathbf{M}$ (cf. Section 2.1) instead of $g_i(\mathbf{P}_l)$ and $\widetilde{\mathbf{M}}$ results in an approximated criterion value $\widetilde{Y}^i$. This is, for example, the case in the prediction of results of the forming process, since a reference mesh is used.*

**Remark 6.2.** *When the normalized database $\widehat{\widetilde{\mathbf{M}}}$ according to Equation (5.1) is used, the previously subtracted mean of each row has to be added back in Equation (6.1).*

In order to validate the precision of the accelerated metamodel, the difference between the predicted result $\widetilde{\mathbf{g}}(\mathbf{P})$ and a reference solution $\mathbf{g}_{\text{ref}}(\mathbf{P})$ is computed component by component. In particular, the absolute difference and the relative difference with respect to the reference solution is considered, given by

$$\mathbf{diff}_{\mathrm{abs}} = |\tilde{\mathbf{g}}(\mathbf{P}) - \mathbf{g}_{\mathrm{ref}}(\mathbf{P})|,$$

$$\mathbf{diff}_{\mathrm{rel}} = \begin{cases} \frac{|\tilde{\mathbf{g}}(\mathbf{P}) - \mathbf{g}_{\mathrm{ref}}(\mathbf{P})|}{|\mathbf{g}_{\mathrm{ref}}(\mathbf{P})|}, & \mathbf{g}_{\mathrm{ref}}(\mathbf{P}) \neq 0, \\ \tilde{\mathbf{g}}(\mathbf{P}), & \text{otherwise.} \end{cases} \tag{6.2}$$

Analogously, the local absolute and relative differences in an arbitrary mesh node $N_i$ are computed, and denoted by $\mathbf{diff}_{\mathrm{abs},i}$ and $\mathbf{diff}_{\mathrm{rel},i}$, respectively.

**Remark 6.3.** *In the following, we use the corresponding deterministic simulation result as reference solution, unless stated otherwise. That is, in order to validate the precision in an out-of-sample point, a corresponding simulation run is performed.*

The acceleration of the metamodel proposed is advantageous as soon as the number of evaluations exceeds the number of experiments Nexp to construct the metamodel. This is due to the fact that, in this case, the costs of the evaluation of the original metamodel would exceed the costs of the computation of the SVD, cf. Section 6.5.

Due to the metamodel acceleration introduced, substantially less memory is required. The compression ratio is given by $\frac{\mathrm{Nexp}}{k+1}$, cf. Definition 5.6. Furthermore, the approximation of the behavior of a new design is much faster, dependent on the number $k$ of singular values retained in the model, compared with using the uncompressed database, i.e., $k = \mathrm{Nexp}$.

Hence, the metamodel can deal with high dimensional data efficiently due to the acceleration. In particular, distributions of criteria based on simulation results can be approximated on the entire geometry and in each timestep. Additionally, the deformations of the geometry can be predicted using the complete database $\widetilde{\mathbf{M}}$, which is of high interest, for example, in a crash processing step. Moreover, the evaluation of the metamodel is very fast, so that an evaluation in a high number of sampling points ($\gg$ Nexp) becomes possible in order to compute statistics of the solution by a (quasi-)Monte Carlo approach. The optimal number of sampling points Npred should be large enough and depends on the application considered. The computation of statistics is explained in Section 6.2.

## 6.1.2 Dealing with Deleted Mesh Elements

In order to use the forecast models also for the prediction of the results in a crash processing step, these forecast models must be able to identify possible failures, e.g., the formation of cracks. Thus, we develop the accelerated metamodels presented further in order to achieve a sufficient prediction, even for process simulations in which these special effects arise. In particular, the metamodel developed is able to localize in which region a fracture may occur with high probability.

Deleted mesh elements represent fractures and cracks in the component under loading in crash simulations. To be more specific, an element is deleted, if the criterion damage exceeds a certain threshold locally. The information about possible fractures in the component are given element based and in each timestep. Furthermore, the criteria values corresponding to these deleted elements are set to zero within the simulation.

Using a state-of-the-art metamodel based directly on these criteria, the element values set to zero would lead to unrealistically low nodal values when mapping the element values to nodal values, as described in Section 2.2.3. This would induce inaccurate prediction results locally in the domains in which deleted elements occur in the simulation results used to setup the metamodel. For this reason, it is essential that the metamodel is able to deal with deleted elements in order to predict the values of the criteria accurately and to localize possible failures correctly.

**Remark 6.4.** *Using the element values directly would cause the same problems of underestimating the criteria in critical areas as using approximated nodal values.*

We define the function $\mathrm{del}(\cdot)$ to distinguish if an element is deleted, that is,

$$\mathrm{del}(\mathrm{Element}_i) := \begin{cases} 1, & \mathrm{Element}_i \text{ is deleted,} \\ 0, & \text{otherwise.} \end{cases} \tag{6.3}$$

Given this information from the simulation results, we derive a procedure to deal with deleted elements properly. For this purpose, we make use of the deletion mechanism used within the simulation. As already stated, $\mathrm{Element}_i$ is deleted, if the criterion damage evaluated in this element exceeds a certain threshold. In this case, the corresponding element value $e_i$ is set to zero.

We reverse this procedure. To be more specific, we perform a continuous extension of the corresponding element values instead of setting them to zero. Thus, the resulting element values remain physically meaningful. To be more specific, the corresponding element value is set to an appropriately higher value as it would have been before the element deletion. This new element value is then used within the mapping to nodal values. This procedure ensures that nodal values are not underestimated by deleted, i.e., zero element values.

This continuous extension can be written as

$$e_i := \begin{cases} \widetilde{e}_i, & \mathrm{del}(\mathrm{Element}_l) = 1, \\ e_i, & \text{otherwise,} \end{cases} \tag{6.4}$$

where an appropriate value $\widetilde{e}_i$ is specified dependent on the criterion considered, as follows.

If the criterion is increasing up to a certain threshold before deletion, as, for example, the criteria effective plastic strain (EPS) and damage, the extension $\widetilde{e}_i$ is determined as the value which is a certain percent higher than the maximal element value of the existing elements in the current state, that is,

$$\widetilde{e}_i = (1 + \delta) \left( \max_{l, \text{del}(\text{Element}_l) = 0} c_l \right), \qquad (6.5)$$

where $\delta > 0$ is a constant properly chosen. If the criterion is decreasing up to a certain threshold before deletion, as, for example, the criterion thickness, the extension $\widetilde{e}_i$ is determined as the value which is a certain percent lower than the minimal positive element value of the existing elements in the current state, that is,

$$\widetilde{e}_i = (1 - \delta) \left( \min_{l, \text{del}(\text{Element}_l) = 0, e_l > 0} e_l \right), \qquad (6.6)$$

where $\delta$ is a constant properly chosen. In particular, $\delta$ should be chosen, so that the extension is continuous. Thus, $\delta$ should be small enough, so that a gap between the values of the existing element values and the value of the deleted elements is avoided. Furthermore, $\delta$ should be large enough, so that nodal values exceeding a certain threshold identify nodes belonging to at least one deleted mesh element after the mapping from element to nodal values has been applied.

**Remark 6.5.** *We have observed that determining $\delta = 0.1$, i.e., ten percent of the existing maximal/minimal element value, gives good results in numerical experiments of the crash application considered.*

We determine a threshold $n_{\text{del},k}(i)$ which indicates failures in state $\text{ts} = i$ of the simulation corresponding to sampling point $\mathbf{P}_k$ as

$$n_{\text{del},k}(i) := \delta_n \left( \max_{l, \text{del}(\text{Element}_l) = 0} e_l \right), \qquad (6.7)$$

where $\delta_n$ is a constant properly chosen dependent on the application considered.

**Remark 6.6.** *If the criterion is increasing, $\delta_n \leq 1$ is appropriate. For simplicity, we only consider increasing criteria, in the following. In the case of decreasing criteria, the derived formulas are vice versa accordingly.*

Whenever a nodal value $n_j$ exceeds the threshold $n_{\text{del},k}(i)$, this node belongs at least to one failed mesh element. Thus, a possible failure is localized. The extension values $\widetilde{e}_i$ and the threshold $n_{\text{del},k}(i)$ are recomputed in each state.

**Remark 6.7.** *The nodal values are usually smaller than the element values $\widetilde{e}_i$ due to the mapping from element to nodal values. $n_j = \widetilde{e}_i$ holds only if all elements to which this node belongs to, are deleted. Thus, specifying $\delta_n = 1$ usually causes that possible failures are marked in too few nodes and the deletion of single elements cannot be represented properly.*

**Remark 6.8.** *We have observed in numerical experiments of the crash application that stresses are removed locally after a crack is initiated. This leads to clearly lower damage values of the elements in the neighborhood of a failed element as the maximal value. Determining $\delta_n = 0.7$ gives good results in the crash processing step considered in the application. This assumes that neighboring element values are at most 70 percent of the maximal element value.*

The continuous extension of element values derived allows a continuous approximation of the criteria considered by the accelerated RBF metamodel introduced in the previous section. However, the forecast model should also be able to reveal the information gained to localize the regions of possible failure. For this reason, a single threshold to be used within the metamodel has to be determined. This threshold $n_{\mathrm{del},\mathbf{M}}(i)$ is derived from the thresholds $n_{\mathrm{del},l}(i), l = 1, \ldots, \mathrm{Nexp}$, of the simulation results contributing to the setup of the metamodel. We define the threshold used within the metamodel corresponding to the database $\mathbf{M}$ as the median of the contributing thresholds, i.e.,

$$n_{\mathrm{del},\mathbf{M}}(i) := \widehat{\mathbf{Q}}_{0.5}(n_{\mathrm{del},l}(i)), \quad l = 1, \ldots, \mathrm{Nexp}. \qquad (6.8)$$

Therewith, possible outliers within the simulation based thresholds do not contribute, which makes it more robust than using the mean value. Using the minimal or maximal threshold value from the simulations overestimates or underestimates the number of nodes belonging to failed elements, respectively.

Whenever a nodal value predicted by the accelerated metamodel exceeds $n_{\mathrm{del},\mathbf{M}}(i)$, this node is highlighted to belong to a deleted mesh element with high probability. Hence, in this region a possible failure is predicted by the forecast model.

## 6.2   Computation of Statistics

The solution of a stochastic problem is characterized by its probability distribution function. Thus, robustness measures, usually based on the mean of the solution and its standard deviation, have to be taken into account in an optimization task in order to achieve a robust design. A survey on robustness measures can be found, for example, in [10]. If the function investigated is nonlinear and the resulting distribution function is asymmetric, the consideration of the mean and standard deviation as robustness measures will lead to wrong implications. In this case, suitable robustness measures take the probability distribution of the solution around the median into account. For example, it has been suggested recently in [95] to use the median as measure of central tendency and the difference between the median and an upper $q$-quantile as measure of dispersion.

In order to approximate the probability distribution function of the criterion considered, the extended accelerated metamodel is evaluated with a (quasi)-Monte Carlo (MC) method. This procedure provides an estimator of the statistics aimed at on the entire mesh. A (quasi)-MC evaluation is suitable, since the prediction of out-of-sample points is very fast. In detail, due to the acceleration of the RBF metamodel introduced, its evaluation in a large number of points becomes possible. Thus, we compute $q$-quantiles in each mesh node with $q \in [0.05, 0.97]$.

**Remark 6.9.** *The tails of the distribution representing the extreme small or high quantiles, for example, $q = 0.99$, are not investigated in this thesis. As already stated, these quantiles are of interest only in a reliability analysis, which is beyond our scope.*

To compute statistics of an arbitrary processing step $\mathbf{g}$, any method can be used which is based only on samples of the parameters. However, the estimation of a $q$-quantile (cf. Section 2.2.1) as a weighted sum of the order statistics usually requires a high number of samples (i.e., Npred $\gg 1000$), see, e.g., [27]. Hence, using a straightforward approach all these samples have to be stored. This leads to an enormous memory requirement, given by the number of sampling points times the number of mesh nodes.

To avoid storing all samples in memory, we use the $P^2$-algorithm for dynamic calculation of quantiles [55] instead of the order statistics in combination with the evaluation of the accelerated RBF metamodel. The procedure to compute an arbitrary $q$-quantile for a criterion considered is given in Algorithm 6.1. In particular, the algorithm describes the acceleration of the RBF metamodel with a SVD in Lines 1, 2 and 6, cf. Section 6.1.1. Moreover, the algorithm describes the usage of this forecast model for the approximation of an arbitrary $q$−quantile of the criterion with the $P^2$-algorithm $P_q^2$.

**Remark 6.10.** *If the database $\widetilde{\mathbf{M}}$ including several timesteps and criteria is used in the setup of the metamodel (cf. Section 6.1.1), the quantiles of the corresponding deformations will be computed as well. If only the distribution function of a criterion is aimed at, the database $\mathbf{M}$ corresponding to this criterion will be used in order to save memory and computational time.*

---

**Algorithm 6.1** Approximation of a $q$–quantile with the accelerated RBF metamodel

---

**Require:** set of sampling points $S := \{\mathbf{P}_1, \ldots, \mathbf{P}_{\text{Nexp}}\}$, corresponding processed database $\mathbf{M}$, $\varepsilon_{\text{SVD,abs}}$, $q \in [0.05, 0.97]$

1: $[\mathbf{U}, \boldsymbol{\Lambda}, \mathbf{V}^T, k] \leftarrow SVD(\mathbf{M}, \varepsilon_{\text{SVD,abs}})$
2: $\{w_l\}_{l=1,\ldots,\text{Nexp}} \leftarrow \text{metamodel}(S, \mathbf{V}^T)$
3: $\{\mathbf{P}_j\}_{j=1,\ldots,\text{Npred}} \leftarrow \text{MC-DoE}(\mathbf{P}, \sigma_{\mathbf{P}})$
4: **for** $j = 1 \to \text{Npred}$ **do**
5:     **for** $i = 1 \to \text{Nnodes}$ **do**
6:         $\widetilde{g}_i(\mathbf{P}_j) \leftarrow \text{acceleratedMetamodel}(\mathbf{P}_j, \mathbf{w}, \mathbf{U}, \boldsymbol{\Lambda}, \mathbf{V}^T, k)$
7:         $\text{q} - \text{quantile} \leftarrow \text{P}_\text{q}^2(\widetilde{g}_i, q)$
8:     **end for**
9: **end for**

---

The $P^2$-algorithm by Jain and Chlamtac stores only five markers, which are updated as more samples are generated, cf. Line 7 of Algorithm 6.1. The five markers are the minimum, the $q/2$-, $q$- and $(q+1)/2$-quantiles and the maximum. The markers are adjusted with a parabolic prediction ($P^2$) formula. Details on the $P^2$-algorithm can be found in [55].

The number of sampling points required for a sufficient quantile estimation depends on the application considered. A high number of samples (Npred $\gg$ 1000) is usually required using a (quasi-)MC method, see Algorithm 6.1, Line 3. In order to obtain the Npred sampling points, a design of experiments (DoE) according to the probability distribution of the parameters can be used. If the probability distribution is not known, a quasi-MC sampling, assuming a uniform distribution, will be performed within the parameter space.

We approximate the corresponding results in each mesh node with the accelerated metamodel for each sampling point, see Line 6. Note that no additional simulation runs of the processing step are needed. Finally, we update the $P^2$-algorithm estimator of the $q$-quantile (Line 7) in each mesh node. That is, the local distribution function is approximated in each mesh node.

The results of the clustering based on the nonlinearity measure, performed in Section 4.2, can be used to reduce the computational time for the estimation of an arbitrary $q$-quantile. In particular, a high number of sampling points Npred is needed only in the clusters $\text{CL}_a$ belonging to nonlinear influences. In this case, only the functions $\widetilde{g}_i$ with $N_i \in \text{CL}_a$ are evaluated. The remaining clusters reflect regions in which the parameters influence the criterion linearly. Hence, in these clusters, a small number of sampling points is sufficient to obtain an appropriate quantile estimate. In conclusion, the Lines 3 to 9 of the Algorithm 6.1 can be performed with different numbers of samplings according to the assignment of the mesh node to the clusters obtained. This is illustrated in Section 7.3.3.

**Remark 6.11.** *The development of specialized quantile estimators is beyond our scope. The $P^2$-algorithm estimates a q-quantile of the results based on the accelerated metamodel instead of the simulation results directly. Thus, the approximation of the q-quantile can only be as accurate as the metamodel itself. Hence, the application of the $P^2$-algorithm is suitable in this case. That is, it gives a fast and accurate approximation for quantiles with $q \in [0.05, 0.97]$. In order to improve the precision and to reduce the computational effort, specialized quantile estimators useful for bulky data, i.e., $\widetilde{\mathbf{M}}$ with a high number of mesh nodes, and/or timesteps, have been presented recently in [19] under specific assumptions, like monotonous functions.*

Statistics on entire meshes are also analyzed in [125], where scatter shapes, the reduction of random fields by means of a spectral decomposition, like principal component analysis (PCA), and their visualization is described. However, the approximation and propagation of the complete probability distribution for entire process chains is not investigated.

In conclusion, given an input distribution of the random parameters, the fast and accurate forecast models derived enable the approximation of the probability distribution of the criteria considered on the entire mesh. Particularly, the range of variations in the response due to the variations in the parameters can be quantified properly, even for entire process chains, with the new methodology developed.

## 6.3 Propagation of All Relevant Information to the Next Processing Step

The new methodology developed, consisting of the individual components proposed in the previous chapters, enables the propagation of all relevant scatter information to the next processing step. Therewith, not only the history of the process due to a single simulation run, but also the influences due to parameter variations in the process history, can be considered in the next processing step.

The iterative mapping procedure and the usage of the gained information for the next processing step is outlined in the remaining of this section, and presented schematically in Figure 6.2.

We suppose that the prediction quality of the forecast model of the previous processing step, for example, a forming process, is sufficient. The mapping procedure starts by transferring the ensemble of simulation results stored in the database of the previous step to the mesh of the current

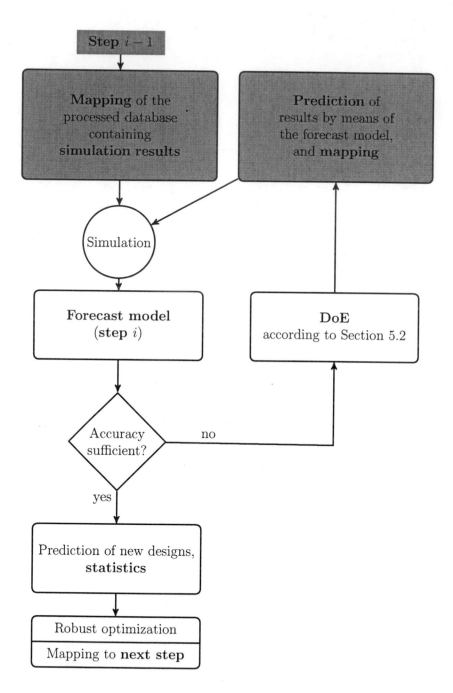

Figure 6.2: Iterative procedure to take all relevant scatter information from the previous processing step ($i-1$, gray) into account in the current processing step ($i$, white).

processing step, for example, a crash process. Then, an ensemble of crash simulation runs is performed. These simulations take the history of the process into account, because the final distributions of the criteria from the forming history, caused by the involved parameter uncertainties, now serve as vector-valued input quantities for the crash simulations. That is, a new database containing an ensemble of simulation runs of the current processing step is generated, which include the history of the process.

Based on this new database a forecast model for the current processing step is set up. If the accuracy of this forecast model, compared with the simulation result as reference solution, is sufficient, the procedure will continue. If the accuracy obtained is not sufficient, the procedure for the iterative extension of the database, introduced in Section 5.2, will be applied. This iteration is based on the forecast model of the previous processing step. For this reason, no additional simulation runs of the previous processing step are required. Instead, the very fast prediction is used.

Then, the forecast model of the previous step is evaluated in these sampling points. The resulting criteria are mapped to the mesh of the current step in order to obtain realistically distributed criteria as input for the simulations. Thereafter, the corresponding simulation runs are performed, and the results are added to the setup of the forecast model. This procedure is iterated until a sufficient prediction accuracy is reached.

After a high-quality forecast model for the current processing step has been created, this model is evaluated according to Algorithm 6.1 in order to approximate the probability distribution function of the criteria of this step.

Finally, either, the approximated quantiles can be used in a robust optimization of the entire process chain, or, all relevant scatter information gained can be mapped and used as input in a further processing step.

This procedure leads to an immense reduction in computational time compared with state-of-the-art quasi-MC simulations for the entire process chain. In conclusion, not only the computational effort in each processing step, but also the overall computational effort to analyze entire process chains involving parameter uncertainties is minimized.

## 6.4 Quality Control

The actual prediction error of the forecast model for a specific out-of-sample point could only be computed, if a simulation is performed in this point. This procedure is not reasonable in each point, since the forecast model should substitute the computationally expensive simulation runs. Thus, we estimate the quality of the forecast model introduced by deriving error bounds for the prediction error in a single processing step and the entire process chain.

The following sources of possible errors have to be taken into account in order to estimate the accuracy of the predicted result considering an arbitrary single processing step $\mathbf{g}$:

- The mappings within a single processing step, cf. Section 2.2.3. A mapping between element values and nodal values is applied, since the methods developed are based on data which is represented as point clouds. Moreover, a mapping between the actual simulation mesh and a reference mesh will become necessary if adaptive mesh refinement is used within the simulation.

- The parameter space dimension reduction, cf. Section 5.1. Only the variations of the most important parameters are taken into account in the forecast model. Non-influential parameters are fixed to their nominal value and not considered as random variables anymore.

- The ensemble compression by means of a SVD of the database, cf. Section 5.3. This transformation maintains the most important information while reducing the computational effort of subsequent analysis steps based on the compressed database.

- The forecast model itself, cf. Section 6.1. An approximation of arbitrary new designs within the parameter space by means of a RBF metamodel is performed.

First, possible errors due to the mapping within a single processing step will be within the mesh discretization error $\mathcal{O}(h^d)$, where $h$ is the mesh size and $d$ is the order of the discretization, if the mapping function is chosen properly. Specifically, this is the case, if the adaptively refined mesh resulting from the last state of the nominal simulation run is used as reference mesh. Therefore, we can neglect this source of error assuming properly defined mappings.

Second, an error is introduced in the model by the reduction of the parameter space, since we neglect parameter variations which do not influence the response at least to a certain amount. This reduction procedure is based on the parameter classification into total importance classes, cf. Chapter 4. The arising error equals the proportion of the overall resulting variation due to parameter variations which cannot be explained by the reduced model. We denote the actual maximal error due to parameter space dimension reduction, measured in percent, by $\text{err}_{\sigma,\max}$.

Third, the ensemble compression by means of a SVD of the database introduces an error by truncation after a certain number of singular values. This error can be measured in each entry of the matrix by the actual difference between the full matrix and its rank-$k$ approximation given by Equation (3.11). In general, the error expresses the proportion of the overall variance present in the database which is not retained in the approximation. We denote the maximal error over all entries due to ensemble compression by means of a SVD, given in percent, by $\text{err}_{\text{SVD,max}}$.

Finally, the forecast model itself is an approximation method, which introduces an additional approximation error. The maximal absolute approximation error is denoted by $\text{err}_{\text{RBF,max}}$.

**Error Analysis for a Single Processing Step**

With the above definitions, we can formulate the following statement.

**Theorem 6.12** (Prediction error in a single processing step). *Let a certain processing step be described in an arbitrary mesh node by a function $g_i$. The processing step on the entire mesh is given by the function vector $\mathbf{g} = [g_1, \ldots, g_{\text{Nnodes}}]^T$. Let the forecast model of the processing step be denoted by $\widetilde{g}_i$, and $\widetilde{\mathbf{g}} = [\widetilde{g}_1, \ldots, \widetilde{g}_{\text{Nnodes}}]^T$, respectively. Let $\text{err}_{\sigma,\text{max}}$ be the actual maximal error over all mesh nodes introduced by the reduction of the parameter space according to Algorithm 5.1, given in percent of the overall variation. Let $\text{err}_{\text{SVD,max}}$ be the maximal error over all mesh nodes introduced by the ensemble compression by means of a SVD, given in percent of the overall variation. Let $\text{var}_{\mathbf{g},\text{max}}$ denote the maximal resulting variation in the response $\mathbf{g}$ according to parameter variations within the parameter space given. Moreover, let $\text{err}_{\text{RBF,max}}$ denote the maximal absolute approximation error of the RBF metamodel over all mesh nodes. Furthermore, we suppose that the problem considered is well-posed. Specifically, we suppose that very small variations in the parameters also have minimal impact on the response. That is, the underlying process is numerically stable. Then, the maximal absolute error in the prediction of a single processing step, i.e., the maximal absolute error in an arbitrary mesh node, can be estimated as*

$$\|\mathbf{g}(\mathbf{P}) - \widetilde{\mathbf{g}}(\mathbf{P})\|_\infty$$
$$\leq \left(1 - (1 - \text{err}_{\sigma,\text{max}})(1 - \text{err}_{\text{SVD,max}})\right) \text{var}_{\mathbf{g},\text{max}} + \text{err}_{\text{RBF,max}}. \quad (6.9)$$

*Proof.* To estimate the overall prediction error, we investigate the different components of the PRO-CHAIN methodology which are performed in order to predict the result of a single processing step. First, the parameter space is reduced. We obtain the proportion of the overall variation which can be explained by the reduced model as $1 - \text{err}_{\sigma,\text{max}}$, because $\text{err}_{\sigma,\text{max}}$ is the actual maximal percentage error over all mesh nodes introduced by the reduction of the parameter space. By multiplying the percentage error with the overall variation, we obtain the absolute error.

The overall variation in the response $\mathbf{g}$ can be estimated with the maximum of the local total sensitivities over all mesh nodes. Recall that the local total sensitivity of parameter $P^j$ in mesh node $N_i$ describes the variation in $g_i$ due to a variation of $P^j$, and is given by $\mathrm{TI}_{ij}$ (cf. Section 4.1.3, Equation (4.24)). We denote the estimator of the overall variation, based on the total sensitivities, by $\mathrm{sens}_{\mathbf{g},\max}$. It is computed as

$$\mathrm{sens}_{\mathbf{g},\max} := \max_{i=1,\ldots,\mathrm{Nnodes}} \left( \sum_{j=1}^{\mathrm{Npar}} \mathrm{TI}_{ij} \right). \tag{6.10}$$

Hence, the maximal absolute error due to the parameter space reduction is given by $\mathrm{err}_{\sigma,\max}\mathrm{sens}_{\mathbf{g},\max}$.

Next, a lossy compression of the entire database $\mathbf{M}$ by means of a SVD is performed, cf. Section 5.3. Analogous to the previous considerations, we obtain the proportion of the overall variation of the database which can be explained by the reduced model, given by the rank-$k$ approximation of the database, as $1 - \mathrm{err}_{\mathrm{SVD},\max}$.

The database contains an ensemble of responses, that is, realizations of the processing step (cf. Chapter 5). Therefore, the maximal deviation from the mean in a row of the database is also an estimator of the maximal variation in the response. We denote this estimator of the overall variation by $\mathrm{variance}_{\mathbf{M}_i,\max}$ and compute it as

$$\mathrm{variance}_{\mathbf{M}_i,\max} := \max_{i=1,\ldots,\mathrm{Nnodes}} \left( \sqrt{(\mathbf{Var}[\mathbf{M}_i])} \right), \tag{6.11}$$

where $\mathbf{M}_i = [M_{i1}, \ldots, M_{i\mathrm{Nexp}}]$ denotes the $i$-th row of the database.

Hence, the maximal absolute error due to the ensemble compression is given by $\mathrm{err}_{\mathrm{SVD},\max}\mathrm{variance}_{\mathbf{M}_i,\max}$.

Bringing this two sources of errors together, a common estimator of the overall variation in the response has to be determined. The maximum of the two derived estimators serve as a properly defined estimator of the maximal variation in the response $\mathbf{g}$. That is, we obtain

$$\max\left(\mathrm{sens}_{\mathbf{g},\max}, \mathrm{variance}_{\mathbf{M}_i,\max}\right) \leq \mathrm{var}_{\mathbf{g},\max}, \tag{6.12}$$

since $\mathrm{var}_{\mathbf{g},\max}$ has been defined as the maximal resulting variation in the response $\mathbf{g}$.

Concluding, $(1 - (1 - \mathrm{err}_{\sigma,\max})(1 - \mathrm{err}_{\mathrm{SVD},\max}))\,\mathrm{var}_{\mathbf{g},\max}$ gives an upper bound of the error due to parameter space reduction and ensemble compression of the database.

Finally, the approximation error of the RBF metamodel is independent of the other errors described above, since the metamodel is already constructed with the processed database. Therewith, we obtain the result desired.                                                                                        $\square$

**Remark 6.13.** *If the response is constant in all mesh nodes, both estimators of the maximal variation in the response will be zero as expected. If the function is nonlinear, the estimators of the variation might differ in the order of magnitude. In this case, the estimation of the total variation by Equation (6.12) may overestimates the actual prediction error given by Equation (6.9).*

The actual errors arising within the different PRO-CHAIN components can usually not be derived analytically, or measured directly. For example, the prediction error within the metamodel can only be computed in the sampling points. In the out-of-sample points, an estimator of the actual error has to be derived. Moreover, the parameter space dimension reduction is based on the accumulated total sensitivities over all mesh nodes. Thus, the maximal error in an arbitrary mesh node, which is needed to compute the error bound derived, cannot be measured without additional computational effort.

We derive a weaker bound of the error, predicting the results of a single processing step, as direct conclusion from the above derivations

**Corollary 6.14** (Estimator of the averaged error). *Suppose that the processing step* $\mathbf{g}$ *and its approximation function* $\widetilde{\mathbf{g}}$ *are given as in Theorem 6.12. Let* $var_{\mathbf{g}}$ *be the averaged variation in the response. Let* $\mathrm{tol}(\widetilde{\mathbf{g}})$ *be an estimator of the average approximation error in the RBF metamodel. Suppose that the problem considered is well-posed. Then, the averaged error over all mesh nodes in an arbitrary point* $\mathbf{P}$ *can be estimated as*

$$\frac{1}{\mathrm{Nnodes}} \|\mathbf{g}(\mathbf{P}) - \widetilde{\mathbf{g}}(\mathbf{P})\|_1$$
$$\leq (1 - (1 - \varepsilon_\sigma)(1 - \varepsilon_{SVD}))\, var_{\mathbf{g}} + \mathrm{tol}(\widetilde{\mathbf{g}}). \qquad (6.13)$$

*Proof.* We use the error estimators of the several PRO-CHAIN components, already derived in Chapter 5. First, the parameter space dimension is reduced according to Algorithm 5.1 by selecting the parameters with a strong influence, which should be retained in the model. This is iterated until a user specified accuracy $1 - \varepsilon_\sigma$ is achieved. This threshold depends on the accumulated total sensitivities. Hence, it expresses the proportion of the overall variation which should be explained on average by the reduced model.

Second, the database is compressed by a SVD. The arising squared average prediction error by using the rank-$k$ approximation of the database is given by the sum of squares of the omitted singular values (cf. Section 3.3.2). Several error estimators of the approximation error due to a SVD have been discussed in Section 5.3. For example, the estimator $\mathrm{err}_2(k)$ represents the average error over all experiments. Then, the number of singular values $k$

retained in the model has been chosen, so that $\text{err}_j(k) < \varepsilon_{\text{SVD,abs}}$, for a specified $j$. The relative average error $\varepsilon_{\text{SVD}}$ is obtained by multiplying $\varepsilon_{\text{SVD,abs}}$ with the average variation present in the database. Hence, $1 - \varepsilon_{\text{SVD}}$ is the proportion of overall variation in the model which can be still explained on average by the rank-$k$ approximation.

In conclusion, the average prediction error, measured in percent, is given by $(1 - (1 - \varepsilon_\sigma)(1 - \varepsilon_{\text{SVD}}))$. Multiplying this with the averaged variation in the response leads to the average absolute prediction error. The average variation in the response can be measured, as outlined in the proof of Theorem 6.12 using the average instead of the maximum.

Finally, the average prediction error in the approximation with the RBF metamodel has to be measured. Since this error cannot be computed, we use the metamodel tolerance as indicator for the actual error. The tolerance can be measured in each arbitrary out-of-sample point with Equation (3.21). Thus, in oder to obtain an average approximation error in an arbitrary point, we evaluate the forecast model in Npred = 100000 random points and compute the average over the corresponding model tolerances, that is,

$$\text{tol}(\widetilde{\mathbf{g}}) := \frac{1}{\text{Npred}} \sum_{i=1}^{\text{Npred}} \text{tol}(\mathbf{P}_i). \tag{6.14}$$

Therewith, we obtain the result desired.  □

This bound of the averaged error derived can be estimated without additional computational effort, since the different components of the PROCHAIN methodology already compute the different error estimators used. Thus, it can be applied directly to industrially relevant applications, as we show in Chapter 7.

### Error Analysis for the Entire Process Chain

With the above definitions and the error bound derived for a single processing step, we can formulate the following statement considering the entire process chain.

**Theorem 6.15** (Prediction error in entire process chain). *Let the first and second processing step be described by the function vectors $\mathbf{g}_1$ and $\mathbf{g}_2$, respectively. Let the corresponding forecast models be denoted by $\widetilde{\mathbf{g}}_1$ and $\widetilde{\mathbf{g}}_2$, respectively. Let $D$ be a neighborhood of $g_{1,i}(\mathbf{P}_{\text{nom}})$ sufficiently large, so that for each mesh node $g_{1,i}(\mathbf{P}), \widetilde{g}_{1,i}(\mathbf{P}) \in D$. Suppose that the function describing the second processing step is locally Lipschitz continuous in this neighborhood $D$ in each component $g_{2,i}$ with Lipschitz constant $L_i$. Furthermore, we suppose that the problem considered is well-posed, specifically, that the underlying process is numerically stable. Let the maximal absolute* *prediction error in the first processing step be given by Equation (6.9) and denoted by the constant $c_0$. Let $\mathbf{A}_1$ and $\mathbf{A}_2$ be different databases used to generate a forecast model for the second processing step. In particular, let $\mathbf{A}_1$ denote a database containing only mapped simulation results of the first processing step. Let $\mathbf{A}_2$ denote a database containing only mapped, predicted results of the first processing step by means of the forecast model $\widetilde{\mathbf{g}}_1$. Let $c_1$ and $c_2$ denote the single prediction error of the second processing step corresponding to the databases $\mathbf{A}_1, \mathbf{A}_2$, respectively, given by Equation (6.9). Then, the maximal absolute error $err_{max,abs}(\mathbf{g}_2 \circ \mathbf{g}_1)$ in the prediction of the process chain $\mathbf{g}_2 \circ \mathbf{g}_1$ is bounded by*

$$c_1 \leq err_{max,abs}(\mathbf{g}_2 \circ \mathbf{g}_1) \leq Lc_0 + c_2, \tag{6.15}$$

*with $L = \max_{i=1}^{\text{Nnodes}} L_i$.*

*Proof.* The maximal absolute prediction error in a single processing step is derived in Theorem 6.12. Hence, the error within the prediction of the first processing step is bounded by

$$\|\mathbf{g}_1(\mathbf{P}) - \widetilde{\mathbf{g}}_1(\mathbf{P})\|_\infty$$
$$\leq (1 - (1 - err_{\sigma,max})(1 - err_{SVD,max}))\,var_{\mathbf{g}_1,max} + err_{RBF,max} =: c_0. \tag{6.16}$$

In the following, we investigate the best and worst case scenario. The forecast model is generated based on the extended database, as explained in Section 5.2.

**Best Case Analysis:** In the best case, the number of available simulation results of the first processing step is sufficient in order to generate a forecast model of the subsequent step. In this case, only simulation results are mapped, no predicted results of the first step are used. The input parameters are given in each mesh node by the function values $g_{1,i}(\mathbf{P}) \in D$. In this case, the Theorem 6.12 is applicable, too. Thus, the prediction error of the composition of both processing steps is bounded by

$$\|\mathbf{g}_2\left(\mathbf{g}_1(\mathbf{P})\right) - \widetilde{\mathbf{g}}_2\left(\mathbf{g}_1(\mathbf{P})\right)\|_\infty$$
$$\leq (1 - (1 - err_{\sigma,max,\mathbf{A}_1})(1 - err_{SVD,max,\mathbf{A}_1}))\,var_{\mathbf{g}_2,max,\mathbf{A}_1}$$
$$+ err_{RBF,max,\mathbf{A}_1} =: c_1. \tag{6.17}$$

**Worst Case Analysis:** In the worst case, no simulation results would be available. In this case, all samples required are predicted results by means of the forecast model of the first processing step. In this case, the prediction error in the second processing step can be estimated as

$$\|\mathbf{g}_2\left(\widetilde{\mathbf{g}}_1(\mathbf{P})\right) - \widetilde{\mathbf{g}}_2\left(\widetilde{\mathbf{g}}_1(\mathbf{P})\right)\|_\infty$$
$$\leq (1 - (1 - err_{\sigma,max,\mathbf{A}_2})(1 - err_{SVD,max,\mathbf{A}_2}))\,var_{\mathbf{g}_2,max,\mathbf{A}_2}$$
$$+ err_{RBF,max,\mathbf{A}_2} =: c_2. \tag{6.18}$$

Using the Lipschitz condition of the second processing step, we obtain

$$
\begin{aligned}
&\|\mathbf{g}_2\left(\mathbf{g}_1(\mathbf{P})\right) - \widetilde{\mathbf{g}}_2\left(\widetilde{\mathbf{g}}_1(\mathbf{P})\right)\|_\infty \\
&\leq \|\mathbf{g}_2\left(\mathbf{g}_1(\mathbf{P})\right) - \mathbf{g}_2\left(\widetilde{\mathbf{g}}_1(\mathbf{P})\right)\|_\infty + \|\mathbf{g}_2\left(\widetilde{\mathbf{g}}_1(\mathbf{P})\right) - \widetilde{\mathbf{g}}_2\left(\widetilde{\mathbf{g}}_1(\mathbf{P})\right)\|_\infty \\
&\underset{(6.18)}{\leq} L\|\mathbf{g}_1(\mathbf{P}) - \widetilde{\mathbf{g}}_1(\mathbf{P})\|_\infty + c_2
\end{aligned}
$$

$$
\underset{(6.16)}{\leq} Lc_0 + c_2. \tag{6.19}
$$

The extension procedure of the database first uses all simulation results available. If the forecast accuracy is not sufficient, predicted results will be added providing that the forecast quality of the first processing step is sufficient. In this case, the database contains a mixture of simulation results and predicted results as input for the second processing step. Thus, the maximal absolute prediction error in the process chain $\mathbf{g}_2 \circ \mathbf{g}_1$ is given by Equations (6.18), (6.19), i.e.,

$$
c_1 \leq \mathrm{err}_{\mathrm{max,abs}}(\mathbf{g}_2 \circ \mathbf{g}_1) \leq Lc_0 + c_2, \tag{6.20}
$$

which is the result desired.                                                    □

**Remark 6.16.** *The worst case scenario does not occur in practice. This is due to the fact that the extension procedure described in Section 5.2 specifies that the available simulation results to generate the forecast model of the first processing step are mapped to the second processing step in any case, in order to improve the forecast quality.*

**Remark 6.17.** *The parameter classification procedure requires that the function* $\mathbf{g}$ *describing the processing step considered is at least continuously differentiable in a neighborhood of the nominal parameter vector. Thus, it follows that* $\mathbf{g}$ *is locally Lipschitz continuous around the nominal parameter vector. In summary, the assumption of Theorem 6.15 does not represent an additional requirement.*

Finally, we quantify the error in the prediction of the probability distribution function in each mesh node. This prediction is based on the approximation of quantiles by means of the $P^2$-algorithm evaluating the forecast model in Npred sampling points (cf. Section 6.2). The average prediction error in the forecast model using Npred samples is estimated by the root mean squared error (RMSE). That is, it follows directly that the RMSE is also bounded, as derived in Theorem 6.15. Thus, the approximation error of the $P^2$ algorithm is added to this prediction error of the forecast model, this gives a bound for the overall average prediction error of the probability distribution in a specific mesh node.

**Remark 6.18.** *The prediction error of process chains consisting of more than two steps can be derived analogously.*

In conclusion, we have derived a bound of the prediction error in each single processing step, as well as, in the composition of several steps, which is used to quantify the prediction error in the resulting distribution function theoretically. The derived bound of the averaged error for a single processing step is computed directly within the applications considered.

## 6.5 Efficiency

We investigate the computational complexity and the memory requirements of the evaluation of the forecast model. The setup of the metamodel used has been described in Section 3.4. Each approximation by means of the metamodel is a matrix vector multiplication, which has the complexity $\mathcal{O}(\text{NnodesNexp})$. If the metamodel is accelerated by a SVD, the computational costs will be reduced to $\mathcal{O}(\text{Nnodes}k)$, where $k$ is the number of singular values retained in the model (cf. Section 6.1.1). Additionally, the SVD of the database has to be computed once, comprising costs of $\mathcal{O}(\text{NnodesNexp}^2)$. Therefore, the accelerated metamodel will be faster than the original one, if the approximation is computed many times for different parameter sets, i.e., Npred $\gg$ Nexp [20]. This is the case, if we use a quasi-MC method using Npred sampling points as described in Section 6.2.

Furthermore, the $P^2$-algorithm is used for the local computation of Nq quantiles in each mesh node. This algorithm updates five markers for each new sample generated. Including the metamodel evaluations, this leads to a computational complexity of $\mathcal{O}(\text{NnodesNpredNq})$, since Npred $\gg$ Nexp holds. Note that the order statistics of the entire set of samples must not be computed, especially, no sorting of this set is required. This makes the dynamic computation of quantiles with the $P^2$-algorithm in combination with the metamodel very efficient. Thus, it is applicable locally on the entire mesh.

When the number of rows of the database is very large, e.g., several criteria and timesteps are considered, the computation of the SVD as well as the evaluation of the metamodel can be accelerated further by a parallelization of these procedures. This is possible, since the local evaluation and the update of the quantile estimators depend only on the current mesh node. That is, blocks of nodes can be evaluated and updated in parallel.

## 6.6   Conclusions

In the previous chapters, we have developed a parameter classification procedure and an iterative processing of the database. By applying these procedures proposed, we obtain a database which reflects all important variations in the criteria considered caused by the parameter uncertainties involved.

The therewith gained information has been used to derive high-quality metamodels. The acceleration of state-of-the-art RBF metamodels enables the efficient prediction of the behavior of new designs locally. Moreover, a procedure has been developed to predict possible failures during a crash process properly, which has not been possible with state-of-the-art metamodels, so far.

In particular, the metamodels include all relevant parameter variations, that is, all variations with a non negligible impact on the criteria considered. The created metamodels enable a fast and accurate prediction of arbitrary out-of-sample points avoiding additional simulation runs. Especially, complete statistical information, given by an approximation of the probability distribution function of the criterion considered in each mesh node, is obtained by evaluating the metamodels generated.

The crash processing step is usually considered separately. At least results of a nominal forming simulation run are considered in the crash processing step, recently. This has become possible by the newest establishment of commercial tools dealing with the technical transfer of vector-valued results between distinct meshes for single simulation runs. Nevertheless, the systematic consideration of parameter uncertainties on the entire mesh in whole process chains is not state-of-the-art. This is particularly for reasons of high computational effort of statistical methods, for example, quasi-MC methods.

With the iterative mapping procedure developed, using fast and high-quality metamodels, it is now possible to perform a comprehensive analysis of the overall influences due to parameter uncertainties in entire process chains, as we demonstrate in the following chapter.

In order to control the approximation quality, we have derived an estimator of the maximal prediction error in a single processing step as well as for the entire process chain theoretically. Furthermore, an estimator of the average prediction error in a new design has been derived. This estimator is computed directly within the methodology developed, without additional computational effort. Hence, it gives a valuable error estimate useful within numerical applications.

In summary, the metamodels include the complete information which is necessary to take the influences of all random parameters involved into account. To be more specific, the prediction of a criterion in an arbitrary parameter vector gives an accurate result under realistic conditions with an user controlled prediction error. Hence, the PRO-CHAIN methodology enables an efficient propagation of all relevant scatter information locally from one processing step to the next allowing a robust optimization of the entire process chain.

# Chapter 7

---

# Benchmarks and Industrial Applications

---

We have developed an innovative methodology, including a parameter classification procedure, and fast and accurate forecast models, which enables the efficient propagation of variations from one processing step to the next. In this chapter, benchmarks and industrial applications are investigated in order to demonstrate the benefits and effectiveness of the new methodology.

First, a numerical comparison between a state-of-the-art method for solving stochastic partial differential equations (sPDEs) and the methodology newly developed is presented in **Section 7.1**. This benchmark shows that the **new methodology gains in computational time compared with a stochastic collocation method** for a model problem. This model problem can be regarded as a single processing step. Thus, the benchmark shows particularly that components of the PRO-CHAIN methodology can be efficiently applied individually.

Second, the fundamental forming example is reviewed in **Section 7.2**, which has been introduced in Section 2.3 and considered in the previous chapters to illustrate individual aspects of the methodology. Now, we compare and discuss the results of several forecast models. This industrially relevant forming application is a good example for the benefit of using the PRO-CHAIN methodology proposed for a single processing step.

Finally, we investigate a forming-to-crash process chain of a B-pillar, which has been provided by Daimler AG, in **Section 7.3**. Due to ever shorter development periods of modern vehicles and an increasing request on passive safety, virtual development methods are more and more established in the automotive industry. However, the quality of crash simulations needs to be improved further. Specifically, all variations arising due to parameter variations within the forming history have to be taken into account to get a realistic prediction, which is still not state-of-the-art. We show that the new methodology can be applied to such **complex process chains from the automotive industry**. In particular, we demonstrate that the PRO-CHAIN methodology newly developed enables the propagation of all relevant variations from one processing step to the next, and thereby, **improves the quality of crash simulations** considerably.

**Remark 7.1.** *In the following, we specify arbitrary out-of-sample points within the parameter space as test points in order to validate the prediction accuracy of the forecast models. In these points $\mathbf{P}_i = [P_i^1, \ldots, P_i^{\mathrm{Npar}}]^T$ corresponding simulation runs are performed, so that the predicted results can be compared with the simulation results. Additionally, we specify several test nodes on the mesh in order to analyze the local process behavior.*

**Remark 7.2.** *We have implemented parts of the new methodology developed in the software DesParO®\* by Fraunhofer SCAI [34], which is used in the analysis of the following applications.*

## 7.1   Numerical Comparison Between the New Methodology and a Stochastic Collocation Method

Specialized methods have been developed to solve sPDEs, see Section 3.5. The PRO-CHAIN methodology proposed provides an alternative approximation approach for the solution of sPDEs with multiple random variables. We compare the new methodology developed with a stochastic collocation method by means of a model problem. In particular, solutions for out-of-sample points are predicted containing a description of the probability distribution function. Both approaches are sampling based. The main differences between the two approaches are the construction of the sample points and the approximation function itself. The comparison for the model problem considered has already been published in [115] which builds the basis for the remaining of this section. A general overview of the model problem is given in Section 7.1.1. Then, the results of the parameter classification procedure are presented in Section 7.1.2. Finally, the stochastic solution obtained by a collocation method is compared with the solution obtained by the new methodology in Section 7.1.3.

---

\*DesParO® is a registered wordmark of Fraunhofer Institute for Algorithms and Scientific Computing SCAI.

## 7.1.1 Overview of the Model Problem

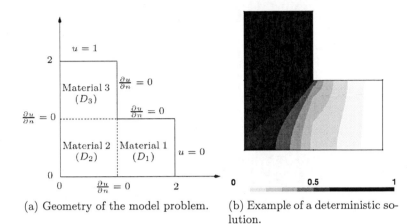

(a) Geometry of the model problem.    (b) Example of a deterministic solution.

Figure 7.1: Geometry of the model problem together with a deterministic solution where all random fields $a_i$ are set to their mean values and $\gamma = 1.0$ (from [115]). The problem is discretized on a mesh with 103479 nodes.

We consider a steady-state diffusion problem (from [115]) with a random diffusion coefficient, defined on a L-shaped domain $\mathcal{D}$, as illustrated in Figure 7.1,

$$-\nabla \cdot (a(\mathbf{X}, u(\mathbf{X}, \mathbf{P}), \mathbf{P})\nabla u(\mathbf{X}, \mathbf{P})) = b(\mathbf{X}) \qquad \mathbf{X} \in \mathcal{D}, \ \mathbf{P} \in \Gamma. \tag{7.1}$$

Dirichlet boundary conditions are imposed on the upper and lower right boundaries, and zero Neumann conditions elsewhere. The right-hand side $b(\mathbf{X})$ is set to one. We model $a(\mathbf{X}, u(\mathbf{X}, \mathbf{P}), \mathbf{P})$ as a nonlinear piecewise random field, consisting of three parts, $a_1$, $a_2$, and $a_3$, defined on the domains $\mathcal{D}_1$, $\mathcal{D}_2$, and $\mathcal{D}_3$, respectively:

$$a(\mathbf{X}, u(\mathbf{X}, \mathbf{P}), \mathbf{P}) = \begin{cases} a_1(\mathbf{X}, \mathbf{P}) + \gamma u^2(\mathbf{X}, \mathbf{P}) & \mathbf{X} \in \mathcal{D}_1, \\ a_2(\mathbf{X}, \mathbf{P}) + \gamma u^2(\mathbf{X}, \mathbf{P}) & \mathbf{X} \in \mathcal{D}_2, \\ a_3(\mathbf{X}, \mathbf{P}) + \gamma u^2(\mathbf{X}, \mathbf{P}) & \mathbf{X} \in \mathcal{D}_3, \end{cases} \tag{7.2}$$

where $\gamma$ is a constant which specifies the nonlinearity. Each component $a_i(\mathbf{X}, \mathbf{P})$, $i = 1, 2, 3$ is assumed to have the form of a truncated Karhunen-Loève (KL) expansion (cf. Section 3.5) with $\mathrm{Npar}_1 = 6$, $\mathrm{Npar}_2 = 7$ and $\mathrm{Npar}_3 = 5$ random variables. Furthermore, all Npar random variables are assumed to be independent, with $\mathrm{Npar} = \mathrm{Npar}_1 + \mathrm{Npar}_2 + \mathrm{Npar}_3 = 18$. We use an exponential covariance function for $a_i$ in the numerical experiments, given by

$$\mathbf{Cov}[\mathbf{X}, \mathbf{X}'] = \sigma^2 \exp\left(-\frac{\|\mathbf{X} - \mathbf{X}'\|_1}{l_c}\right), \tag{7.3}$$

where $l_c$ denotes the correlation length. The mean values of $a_i$ are set to $\mathbf{E}[a_1] = 30$, $\mathbf{E}[a_2] = 5$, and $\mathbf{E}[a_3] = 100$, respectively. We consider two configurations for creating the KL-expansions of $a_i$ in Equation (7.2):

- Npar uniformly distributed random variables on $[-\sqrt{3}, \sqrt{3}]$: covariance function (Equation (7.3)) with correlation lengths $l_{c,1} = 1$, $l_{c,2} = 0.5$ and $l_{c,3} = 1.5$; and variances $\sigma_1^2 = 100$, $\sigma_2^2 = 2.25$, $\sigma_3^2 = 900$.

- Npar standard normally distributed random variables: covariance function (Equation (7.3)) with $l_{c,1} = 1$, $l_{c,2} = 0.5$ and $l_{c,3} = 1.5$; and variances $\sigma_1^2 = 9$, $\sigma_2^2 = 0.25$, $\sigma_3^2 = 100$.

The problem is solved on a spatial finite element (FE) mesh with two different resolutions, Nnodes = 12154 and Nnodes = 103479. In each case, we set the constant $\gamma$ for varying the nonlinearity (Equation (7.2)) to 1.0 and 100.0. An example of a deterministic solution is shown in Figure 7.1.

A sparse grid of Clenshaw-Curtis or Gauss points (cf. Section 3.1), denoted by 'Scc' or 'Sg', respectively, is used within the stochastic collocation method. The level of the method is indicated by the number following 'Scc' and 'Sg'.

Since the exact solution $u$ of the sPDE is not known, we base the error computation on a reference solution $u_{\text{ref}}$, which is obtained as a high-order stochastic collocation solution, since this method is a commonly applied state-of-the-art interpolation approach in the context of sPDEs.

## 7.1.2   Results of the Parameter Classification

In order to reduce the parameter space, and, thus, the computational complexity, we apply the parameter classification procedure developed (cf. Chapter 4) to the model problem. To construct the database, a star-point design of experiments (DoE), as described in Section 3.1, with 2Npar + 1 = 37 sampling points and corresponding deterministic solver runs is performed.

The resulting accumulated sensitivity measures are exemplified in Figure 7.2 by the case of 18 Gaussian distributed parameters and 103479 nodes used for discretization.

**Remark 7.3.** *The accumulated sensitivity measures have been computed with the $L_2$-norm in [115]. The advantage of the expected value used in this thesis is that the sensitivities are given directly in the magnitude of the criterion (cf. Section 4.1.1). However, the order and relative influences of the parameters remain the same.*

The sensitivity analysis gives similar results in the cases of uniform and Gaussian distributions for different resolutions of the mesh. If we use only 12154 spatial degrees of freedom the indicator for nonlinearity is higher than

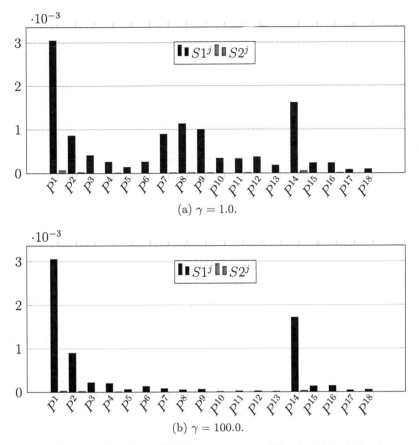

(a) $\gamma = 1.0$.

(b) $\gamma = 100.0$.

Figure 7.2: Accumulated sensitivity measures. The height of the bars correspond to the linear influences ($S1^j$) and the nonlinear influences ($S2^j$) of the Gaussian distributed parameters on the solution.

with a high resolution, nevertheless all parameters are assigned to NLClass$_2$, i.e., they behave linear on average. The ranking of the parameters remains the same for both resolutions, for details refer to [115].

The variation of $\gamma$ in the model problem leads to differences in the ranking and the amount of influence of the parameters on the solution. In detail, Figure 7.2 shows that in the case of $\gamma = 1.0$ the parameters $P^1, P^2, P^7, P^8, P^9$ and $P^{14}$ are the most influencing ones. The corresponding parameter space can be reduced to six instead of 18 parameters, so that the database is reduced to one-third of its original size. In the case of $\gamma = 100.0$ the influence of the parameters at the first few ranks increases, so that the parameter space can be reduced to only three parameters, namely $P^1, P^2$ and $P^{14}$. The indicator of nonlinearity shows no great nonlinear influence of any parameter. In the cases where the Equation (4.15) is fulfilled, that is, for $P^{10}$ and $P^{13}$ in the case of $\gamma = 100.0$, we find that both the linear as well as the nonlinear influence is almost zero. Therefore, these parameters do almost not influence the solution at all.

To confirm the linear behavior of all random variables, we compute the full Hessian matrix and the measure $D$, given by Equation (4.21), in case of uniformly distributed random parameters and a discretization with 12154 nodes. This requires $4\mathrm{Npar}(\mathrm{Npar} - 1)/2 = 612$ additional simulation runs to compute the approximation of the second derivatives. We obtain

$$|\alpha_{\mathrm{max}}| = 1.05487, \quad D = 3.60796 > 0, \tag{7.4}$$

with $\sigma = \sqrt{3}$. This proves the linear behavior of all random variables, at least up to second order approximations.

## 7.1.3  Comparison Between the Results of a Collocation Method and the Accelerated Metamodel Approach

We approximate the probability distribution function by the computation of quantiles using the metamodel (Algorithm 6.1) for several configurations of the model problem given by Equation (7.1). In the following, two cases are presented in detail. In the first case, 18 uniformly distributed random variables on a mesh with 12154 nodes and nonlinearity coefficient $\gamma = 1.0$ are used. Here, the parameter space is reduced to six random variables as a result of the previous parameter classification procedure described above. As a second example, we analyze the case of 18 Gaussian distributed random variables on a mesh with 103479 nodes and nonlinearity coefficient $\gamma = 100.0$. In this case, the parameter space is reduced to three random variables. This second example will be denoted by `Coll-18-Gauss-100`, if it is computed with the stochastic collocation method. It will be denoted by `accRBF-3-Gauss-100`, if it is computed with the new methodology proposed.

In the following, we always use the metamodels with reduced parameter space, unless stated otherwise. A further acceleration by means of a singular value decomposition (SVD) is not needed due to the small database. Thus, the full database is used.

To evaluate the metamodel, we create a new DoE with $\mathrm{Npred} = 2000$ random sampling points $\mathbf{P}_k, k = 1, \ldots, \mathrm{Npred}$. The deterministic solutions of the sPDE according to these sampling points are computed by approximation with the metamodel using Equation (6.1). As an example, the deterministic result for one single sample point $\mathbf{P}_1 = [P^1 = 0.2, P^2 = 0.4, P^{14} = 0.8, P^i = 0.5 \quad \forall i \backslash \{1, 2, 14\}]^T$ is shown in Figure 7.3 on the left. On the right-hand side, the absolute difference between the approximation

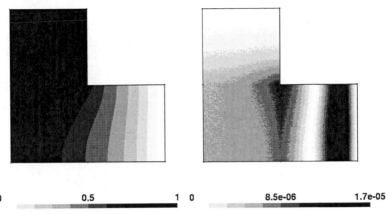

0       0.5       1    0       8.5e-06       1.7e-05

(a) Deterministic solution for sampling point $\mathbf{P}_1$.

(b) Absolute difference between the approximation result in $\mathbf{P}_1$ and the deterministic solution.

Figure 7.3: Deterministic solution and absolute difference between the approximation result using accRBF-3-Gauss-100 and this deterministic solution.

result in this sampling point and the corresponding deterministic solution is shown. Figure 7.3 shows that the approximation with the accelerated metamodel gives accurate results. To be more specific, the maximal absolute error is $1.72907e-05$ using the reduced parameter space. The maximal relative error is $0.012$ percent of the solution value which appears only in the region where the deterministic solution is approximately zero. The whole set of sample points and corresponding approximation results is used to compute $q$-quantiles with $q \in [0.1, 0.9]$.

We compare stochastic collocation methods with the new PRO-CHAIN methodology with respect to the number of deterministic partial differential equations (PDEs) to be solved, the computational time, and accuracy of the computed quantiles for the model problem considered.

The collocation methods use a sparse grid to construct the collocation points, whereas the PRO-CHAIN methodology starts with a star-point DoE resulting in a different number of sampling points, see Table 7.1, and, therewith, a different number of deterministic PDEs to solve.

Details of the different components of the methods and their computational time are listed in the Tables 7.2 and 7.3. All computations are performed on a standard 3GHz Linux PC. Beside the different computational time for the simulations (solution of deterministic PDEs), the different times for constructing the sampling points for the statistics stand out,

| Random variables | Scc1 | Scc2 | Sg1 | Sg2 | PRO-CHAIN |
|---:|---:|---:|---:|---:|---:|
| 18 | 37 | 685 | 37 | 757 | 37 |
| 6 | 13 | 85 | 13 | 109 | 13 |
| 3 | 7 | 25 | 7 | 37 | 7 |

Table 7.1: Number of required sampling points for different numbers of random variables.

| Runtime for | Sg1 | Sg2 |
|---:|---:|---:|
| Construction collocation points | 0.0000E+00 | 0.0000E+00 |
| Simulations | 2.1710E+03 | 4.2733E+04 |
| Construction DoE | 1.9190E+01 | 4.1629E+02 |
| 0.68-quantile | 1.0220E+01 | 1.0440E+01 |
| **Overall** | **2.2004E+03** | **4.3160E+04** |

Table 7.2: Detailed overview of runtime in seconds to compute the 0.68-quantile with the stochastic collocation methods Sg1 and Sg2 for the model problem `Coll-18-Gauss-100`.

| Runtime for | accRBF-3-Gauss-100 |
|---:|:---:|
| Construction DoE | 0.0000E+00 |
| Simulations | 2.1710E+03 |
| Parameter classification | 3.5040E+00 |
| Construction DoE | 0.0000E+00 |
| SVD | 1.3700E+00 |
| 0.68-quantile | 1.1210E+01 |
| **Overall** | **2.1871E+03** |

Table 7.3: Detailed overview of runtime in seconds to compute the 0.68-quantile with the accelerated metamodel `accRBF-3-Gauss-100`.

when comparing the collocation method with the new methodology. The collocation method uses Npred = 2000 sampling points according to the probability distribution of the random variables to reconstruct the stochastic solution given by Equation (3.28). The same number of sampling points Npred = 2000 is also used for the evaluation of the accelerated metamodel. The tables demonstrate that the parameter classification procedure is very fast, although it computes measures in each mesh node. The time for computing the 0.68-quantile using the metamodel contains the time for the approximation of the solution in all sampling points and the update of the quantile estimator with the $P^2$-algorithm according to Algorithm 6.1.

**Remark 7.4.** *The same algorithm for the computation of quantiles, namely the $P^2$-algorithm, is used in both methods to get a meaningful comparison.*

The higher overall runtime to compute statistics with the collocation method is only due to the reconstruction of the stochastic solution. Altogether, the runtime using the new methodology is all about the same than using the collocation method with Gauss points, level one (Sg1), and a lot faster than using the collocation method with Gauss points, level two (Sg2). Note that this comparison only illustrates where the computational effort of each method lies on. Each method may be accelerated by parallelizing components of the method.

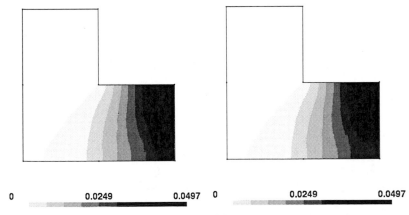

(a) 0.68-quantile, Metamodel vs Sg1.  (b) 0.68-quantile, Metamodel vs Sg2.

Figure 7.4: Relative difference between 0.68-quantile obtained with collocation methods for the model problem `Coll-18-Gauss-100` and the accelerated metamodel using `accRBF-3-Gauss-100`.

We compare the metamodel solution with the reference solutions $u_{ref} = Sg1$ and $u_{ref} = Sg2$. However, note that the computed quantile with

$Sg1$ or $Sg2$ is the exact solution of the constructed polynomial solution approximation, but not the exact solution of the sPDE. We suppose that the higher order method $Sg2$ gives a more accurate solution. Figure 7.4 shows the relative differences (Equation (6.2)) between the 0.68-quantile computed with the collocation method and computed with the accelerated metamodel. Differences are particularly obvious in the domain where the solution of the sPDE is almost zero. The differences are smaller when compared with $Sg2$. The differences are of course influenced by the number of sampling points. For comparison each method uses 2000 sampling points, but an evaluation of the metamodel with more samples may even improve the result.

In summary, the metamodel performs as fast as the collocation $Sg1$ method for the model problem considered. But, its solution is nearer to the solution obtained by $Sg2$. Therefore, the numerical results illustrate that the new methodology is an efficient and accurate alternative to the collocation method to compute statistics of a sPDE containing independent random variables. It is especially advantageous when the probability distributions of the random variables are not known a priori.

## 7.2    Forming of a Pan with Secondary Design Elements

The forming of a pan with secondary design elements is an industrially relevant application, which has already been introduced in Section 2.3. It has been considered in the previous chapters to illustrate the individual components of the new methodology developed. In particular, the results of the parameter classification have been discussed in Section 4.1.4.

In the remaining of this section, we aim at predicting the probability distribution function of the criteria effective plastic strain (EPS) and thickness with the methodology proposed. Therefore, we derive appropriate forecast models in Section 7.2.1. After that, we perform the computation of statistics for this example in Section 7.2.2. Moreover, the results are compared with a state-of-the-art quasi-Monte Carlo (MC) solution. Finally, a brief summary of the results and the benefits of the new methodology for this example is given in Section 7.2.3.

### 7.2.1    Forecast Models

In this pan forming example, we aim at predicting the criteria EPS and thickness by means of the accelerated radial basis function (RBF) metamodel introduced in Section 6.1.1. Based on the results of the total importance classes and the subsequent parameter space dimension reduction (cf. Section 5.1), we set up and compare several metamodels in order to predict the behavior of new designs. The models investigated are listed in Table 7.4.

| Model name | Criterion | Parameters | % | Nexp |
|---|---|---|---|---|
| pan-eps-small | EPS | $\mu, n$ | 71 | 21 |
| pan-eps-big | EPS | $R_{90}, \mu, F^H, K, n$ | 97 | 66 |
| pan-thickness-small | thickness | $t, \mu$ | 82 | 21 |

Table 7.4: Overview of different metamodels investigated for the pan forming example. % refers to the average percent of variation which is expected to be explained by each model. Nexp denotes the number of samples used to create the corresponding metamodel.

**Remark 7.5.** *We do not need the continuous extension derived for deleted mesh elements in this example, because the parameter variations considered do not cause a failure during the forming process.*

We proceed in two steps considering the criterion EPS. In the first step, the model `pan-eps-small` is explored which takes only two parameters, namely the friction coefficient $\mu$ and the strain hardening index $n$, into account. This model is expected to explain on average at least 71 percent of the overall variation according to the dimension reduction results presented in Section 5.1. In the second step, we compare this model with the model `pan-eps-big` which is based on the union of the first, second, and third total importance class. Hence, in addition to $\mu$ and $n$ the parameters $R_{90}, F^H, K$ are considered resulting in a total of 5 random parameters. This bigger model is expected to explain on average at least 97 percent of the overall variation. Moreover, we estimate an upper bound of the average prediction error for both models according to Corollary 6.14. The average variation in the response is given by the estimator based on the total sensitivities (cf. Section 6.4), i.e., $\mathrm{var}_{\mathbf{g}} = \sum_{j=1}^{\mathrm{Npar}} \mathrm{TI}^j = 0.02537$. Note that the local variation in some nodes is much higher. Hence, using the threshold $\varepsilon_\sigma$ specified as in the parameter space reduction described in Section 5.1 and $\varepsilon_{\mathrm{SVD}} = 0.02$ in both cases, we obtain

$$\frac{1}{\mathrm{Nnodes}} \|\mathbf{g}(\mathbf{P}) - \widetilde{\mathbf{g}}(\mathbf{P})\|_1 \leq \qquad (7.5)$$

$$\begin{cases} (1 - (1 - 0.3)(1 - 0.02))\, 0.02537 + 0.0013 = 0.009, & \texttt{pan-eps-small}, \\ (1 - (1 - 0.05)(1 - 0.02))\, 0.02537 + 0.0022 = 0.0039, & \texttt{pan-eps-big}. \end{cases}$$

These thresholds of the average error derived also indicate the higher accuracy of the model `pan-eps-big`. Note that the model tolerances are rather high. However, these measure the sensitivity of the model on the removal of sampling points. Due to the minimization of the number of sampling points to construct the metamodel, a high sensitivity on these points is expected. This is especially true, if the dimension of the parameter space is high, which also explains the higher model tolerance using `pan-eps-big`. Nevertheless, the actual error can be much smaller. In the following, we proof that both error thresholds are fulfilled, and, that the forecast quality is even much better in the majority of mesh nodes.

With regard to the criterion thickness, we only investigate the model `pan-thickness-small` which is based on the two parameters of the first total importance class, $t$ and $\mu$. This model is expected to explain already 82 percent of the overall variation (cf. Table 4.2 and Section 5.1). We do not set up another metamodel for the criterion thickness, taking more parameters into account, because this forecast model is already very accurate in all regions of interest, as we show in the remaining of this section.

All metamodels are constructed according to Equation (6.1) with detrending using polynomial degree one. Unless stated otherwise, we use the SVD with all modes, that is, $k = \mathrm{Nexp}$, in the accelerated metamodel in order to measure the prediction error due to the metamodeling and the error due to ensemble compression separately.

The number of sampling points is set according to Equation (3.22) with $C = 2$ minimizing the number of simulation runs required. Thus, we have performed a DoE with $\mathrm{Nexp} = 21$ simulation runs using two parameters and $\mathrm{Nexp} = 66$ simulation runs using five parameters. Note that these include the $2\mathrm{Npar} + 1$ simulation runs already performed in the parameter classification, so that the database is extended by $\mathrm{Nexp} - (2\mathrm{Npar} + 1)$ simulations, cf. Section 5.2. A comparison between these DoEs projected in the two dimensional parameter space for the criterion EPS is shown in Figure 7.5.

We specify 16 randomly selected out-of-sample points within the entire parameter space in order to validate the predicted results. These test points are highlighted within the reduced two-dimensional parameter space in Figure 7.5 to compare their locations with the sampling points. As reference solution we use the corresponding simulation results in these test points.

**Remark 7.6.** *The 16 test points represent the full uncertainties considered within the application. That is, they include random variations of all six original parameters involved, cf. Table 2.1.*

Additionally, we have specified ten mesh nodes as test nodes in order to analyze the local process behavior, which are shown in Figure 7.6. The mesh nodes selected are chosen at interesting areas of the pan with respect to the resulting criteria. For example, we identify some mesh nodes at the secondary design elements ($N_4 - N_{10}$), or at the wall and edges of the pan

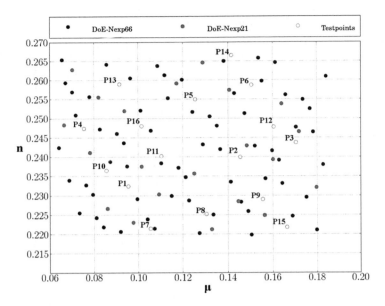

Figure 7.5: DoE with Nexp = 21 and Ncxp — 66 sampling points within the two-dimensional parameter space of `pan-eps-small`. Additionally, 16 out-of-sample points $\mathbf{P}_1, \ldots, \mathbf{P}_{16}$ used as test points for validation are highlighted.

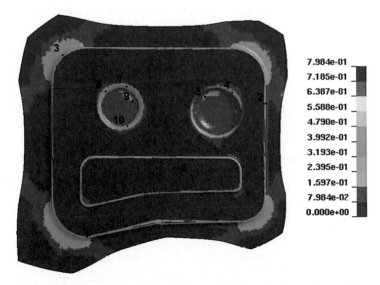

Figure 7.6: Distribution of the criterion EPS resulting from the nominal simulation run with selection of ten mesh nodes, $N_1, \ldots, N_{10}$, in order to analyze the local process behavior.

$(N_1 - N_3)$. These are critical areas in which the material is deeply drawn, which may cause high EPS values, or severe thinning, locally. One test node, $N_6$, is placed within a flat area of a secondary design element. This is a representative of a local region in which the EPS is almost zero.

**Forecast Models To Predict the Criterion EPS**

In the following, we investigate the constructed metamodels in detail, starting with the two forecast models for the criterion EPS. First, the correlation

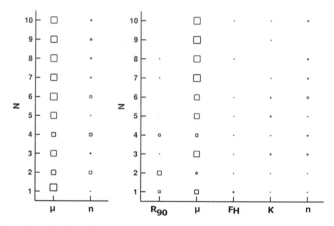

Figure 7.7: Correlation between parameters and the criterion EPS in the ten test nodes for `pan-eps-small` (left) and `pan-eps-big` (right). The size of the square corresponds to the magnitude of the correlation.

between the parameters and the criterion EPS in the ten test nodes specified (cf. Figure 7.6) is analyzed in order to characterize the local process behavior, see Figure 7.7. As expected from the parameter classification results (cf. Table 4.2), the parameter friction has the strongest influence on all test nodes in model `pan-eps-small`. The same dependencies are obvious in `pan-eps-big`. Additionally, higher influences of the anisotropy parameter $R_{90}$ emerge locally on test nodes $N_1, N_2, N_4$, which are not represented in the small model with two parameters.

A more detailed view on the correlation between the friction coefficient and EPS in representative test nodes $N_4, N_7$, and $N_{10}$ is given in the histograms in Figure 7.8. These histograms count the frequency of occurrence of an EPS value evaluating the metamodel `pan-eps-big` in 100000 random points within the parameter space. The small model already shows the same tendency of distribution and is, thus, not shown here. The overall variation of EPS in $N_4$ is very small as expected from the correlation matrix in Figure 7.7. Overall, a non-monotonous, quadratic dependency can be

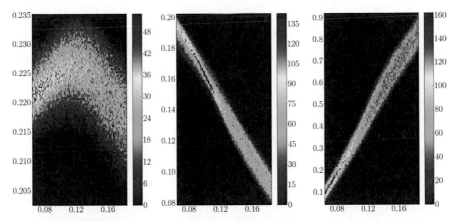

Figure 7.8: Correlation between parameter friction (x-axis) and the criterion EPS (y-axis) in the test nodes $N_4$, $N_7$, and $N_{10}$ (from left to right). The color coding corresponds to the frequency of occurrence of a certain EPS value.

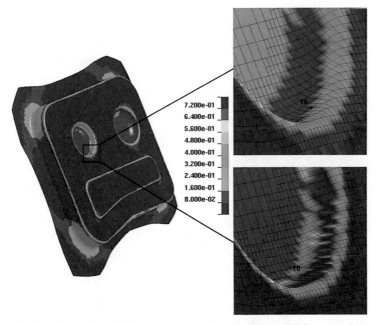

Figure 7.9: High local variation range in the criterion EPS around test node $N_{10}$ (right) resulting from minimal (top) and maximal (bottom) variation of the parameter friction compared with the result of the nominal simulation run (left).

seen with extremum approximately at the center of the parameter space in direction $\mu$. We obtain a strong linear, monotonous decreasing dependency in $N_7$ and a strong linear, monotonous increasing dependency in $N_{10}$, in accordance with the correlation matrix.

The test node $N_{10}$ is located at the wall of one of the secondary design elements. In this area, the friction coefficient has a strong influence and causes a high variation of EPS locally. This is obvious also from the simulation results, see Figure 7.9. If the friction coefficient is small, the resulting EPS value is low, too. If the friction coefficient is very high, the resulting EPS value increases considerably and the sheet is thinned out severely.

A 3-dimensional surface plot of the forecast model is given in Figure 7.10, which confirms the previous observations. The points shown mark the sampling points used to construct the metamodel.

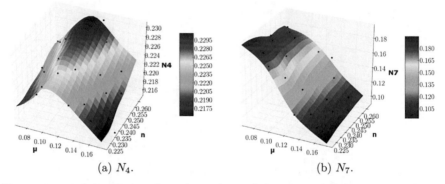

(a) $N_4$.          (b) $N_7$.

Figure 7.10: 3-dimensional surface plots of the metamodel in test nodes $N_4$ and $N_7$ together with the sampling points used for the metamodel setup.

In summary, the investigation of the local correlation between parameters and the criterion EPS points out that the local influence of the parameters considered is very different, even if looking only at the test nodes $N_7 - N_{10}$ at one of the secondary design elements. This confirms that local investigations are required in order to characterize the process behavior sufficiently.

In order to assess the precision of the metamodel, we compute the absolute and relative differences to a reference solution according to Equation (6.2). In particular, the predicted results in the 16 test points using `pan-eps-small` are compared with the corresponding simulation results.

The local absolute and relative prediction errors in the ten test nodes specified for the 16 test points are shown in Figure 7.11. Most of the absolute errors are lower than 0.01 except for $N_8$ and $N_{10}$. The relative errors are

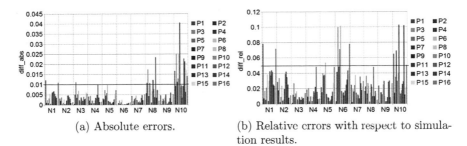

(a) Absolute errors.

(b) Relative errors with respect to simulation results.

Figure 7.11: Local absolute and relative prediction errors in the ten test nodes for the 16 test points using `pan-eps-small`.

rather small, mostly below five percent, except for $N_6$ and $N_{10}$. Specifically, the higher absolute errors in $N_8$ correspond to low relative errors. The higher relative errors in $N_6$ correspond to small absolute errors due to the fact that the EPS value considered itself is almost zero. The higher absolute and relative error in $N_{10}$ can be explained by the high variation of EPS in this node. An adaptive metamodel with more sampling points locally in this area should improve the prediction quality. Nevertheless, evaluating the relative prediction error, the forecast quality of `model-eps-small` is clearly better in all test nodes for all test points than expected from the average estimate of the prediction quality derived in Section 5.1.

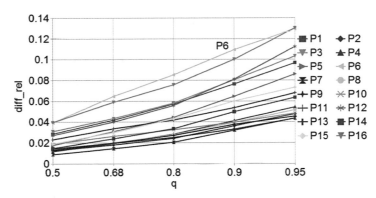

Figure 7.12: Approximated distribution function of relative prediction errors in the 16 test points using `pan-eps-small`.

To verify that this is also the case on the entire geometry, we consider the probability distribution function of the relative prediction errors in the 16 test points, illustrated in Figure 7.12. This distribution is approximated by $q$–quantiles with $q \in [0.5, 0.95]$ computed from the order statistics (cf. Section 2.2.1).

For two test points, we obtain generally higher relative errors compared with the remaining points. The location of these test points within the sampling space is a possible reason for this. Specifically, the test point $\mathbf{P}_6$ is located at the border of the spanned parameter space. In this point, the evaluation is within the valid range, but the metamodel starts to extrapolate. Hence, its evaluation in this point is not expected to give meaningful results.

Another possible reason for higher relative errors is that the model `pan-eps-small` is based only on two parameters explaining at least 70 percent of the variation on average, whereas the test points include variations of all six parameters. Nevertheless, the upper bound of the average absolute prediction error, derived with Equation (7.5), is fulfilled (cf. Figure 7.11a).

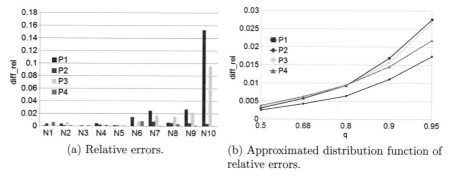

(a) Relative errors.

(b) Approximated distribution function of relative errors.

Figure 7.13: Relative prediction errors in the ten test nodes specified together with their approximated distribution in four test points including only variations of the two relevant parameters $\mu, n$ using `pan-eps-small`.

Thus, we compare the predicted results in four additional test points which only include variations of the two relevant parameters $\mu$ and $n$. All other parameters are set to their nominal value. In these points, the absolute and relative errors are much smaller. The approximated distribution of the relative errors, presented in Figure 7.13, shows that the 0.95-quantile is less than three percent in all four test points considered.

Finally, we compare the results with the model `pan-eps-big` which contains more sampling points, and also takes variations of more parameters into account. The local relative errors in the ten test nodes specified for the 16 test points are shown in Figure 7.14. We observe that most of the relative errors are substantially lower than with the model `pan-eps-small`, except for $N_{10}$. The variation and the differences in this test node have

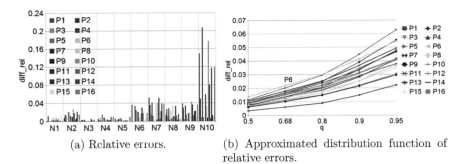

(a) Relative errors.

(b) Approximated distribution function of relative errors.

Figure 7.14: Local relative prediction errors in the ten test nodes and their distribution for the 16 test points using `pan-eps-big`.

increased, but the absolute error in this test node stays less than 0.04. Furthermore, the upper bound of the average absolute prediction error, derived with Equation (7.5), is fulfilled. The higher relative prediction errors in $N_{10}$ indicate that the overall resulting variation in this mesh node cannot be fully explained by the five parameters considered.

The test point $\mathbf{P}_6$ is not located at the border of the sampling space anymore due to the increase of the sampling points. This leads to a significant decrease of the errors in this point, which can be seen in Figure 7.14 compared with Figure 7.12. The approximated distribution function of the relative errors, graphed in Figure 7.14b, confirms the highly improved prediction quality of `pan-eps-big`. Overall, 80 percent of the relative errors are below the expected average error of 3 percent.

The improvement of the prediction with `pan-eps-big` is also clearly visible in the local prediction errors of the entire geometry. Exemplary, the relative errors occurring in $\mathbf{P}_6$ and $\mathbf{P}_9$ are shown in Figure 7.15. The test point $\mathbf{P}_6$ has been at the border of the parameter space using the small model and it lies within the parameter space using the big model. Therefore, much better results are obtained and the errors decrease substantially. The test point $\mathbf{P}_9$ is a representative for the test points positioned in the middle of the parameter space. The results have been good already with the small model. The bigger model improves the results locally. Specifically, the higher errors at the left secondary design element disappear, highlighted in Figure 7.15b. Recalling the results of the parameter classification procedure, this was also expected from the local sum of sensitivities presented in Figure 5.1. This figure also shows that variations in a few mesh nodes at the deep drawn wall and the borders cannot be fully explained by the five parameters considered. This agrees with the higher prediction errors in these nodes, for example, $N_{10}$, as we have already shown in Figure 7.14.

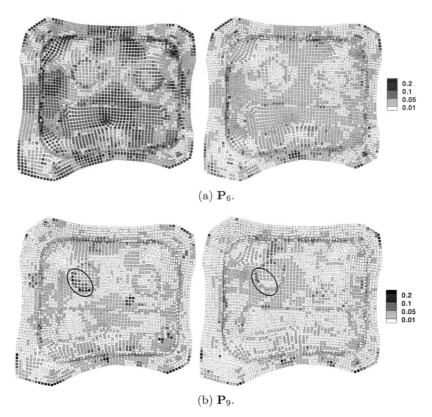

(a) $\mathbf{P}_6$.

(b) $\mathbf{P}_9$.

Figure 7.15: Local relative prediction errors on the entire geometry in two selected test points using `pan-eps-small` (left) and `pan-eps-big` (right).

## Forecast Models To Predict the Criterion Thickness

We briefly summarize the model `pan-thickness-small` to predict the criterion thickness. This model shows linear dependencies in all test nodes, except for $N_7$. Hence, a correlation analysis is not shown for this model. The small model already gives a high accuracy due to the resulting smooth behavior and the dominant influence of the parameter thickness.

The relative errors in the test nodes and the approximated distribution function of the relative errors in all test points is presented in Figure 7.16. The relative errors in all test nodes are less than three percent, except in $N_{10}$, where the relative error is less than five percent. The distribution functions of the relative errors show that all quantile values are less than one percent, except for one test point. Thus, the model is locally already better than the average expected quality derived by the parameter classification procedure in most mesh nodes.

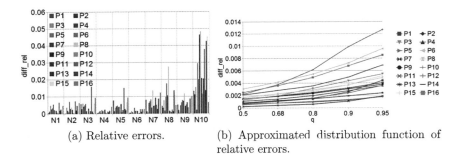

(a) Relative errors.　　(b) Approximated distribution function of relative errors.

Figure 7.16: Relative prediction errors in the ten test nodes and their approximated distribution function for 16 test points using `pan-thick-ness-small`.

## 7.2.2 Computation of Statistics

In the previous section, we have shown that appropriate forecast models to predict the criteria EPS and thickness have been derived. We use these high-quality metamodels to approximate the probability distribution functions of the criteria considered. These can be used in a subsequent robust optimization.

For this purpose, several $q$-quantiles with $q \in [0.1, 0.97]$ are approximated according to Algorithm 6.1. The approximation is based on the metamodels constructed, as listed in Table 7.4. The number of samples is increased from Npred = 500 to Npred = 100000 in order to show the convergence of the quantile estimates. The computational time to estimate the quantiles with the algorithm derived is linear in the number of samples. An estimation of the quantiles in all mesh nodes is very fast. For example, the runtime to estimate nine quantiles in each mesh node using Npred = 1000 samples on a standard 3 GHz Linux PC is approximately 1.03 seconds in this example.

For comparison, we estimate the $q$–quantiles with a quasi-MC simulation based directly on the simulation results. That is, we perform Npred = 1000 simulation runs according to a Halton sequence (cf. Section 3.1). Afterwards, the quantiles required are estimated using the order statistics, see Equation (2.19).

**Remark 7.7.** *The quasi-MC simulation may not provide a good reference solution for the quantiles considered, since the exact stochastic solution is not known, and the number of samples is possibly too small. More simulations could not be performed due to restrictions in computational time.*

The quantiles estimated by the forecast model are compared locally in the test nodes specified (cf. Figure 7.6) with the quantiles estimated by the quasi-MC procedure used. Exemplary, a comparison in the test nodes $N_7$ and $N_{10}$ is given in Figure 7.17. The figure points out that the estimated

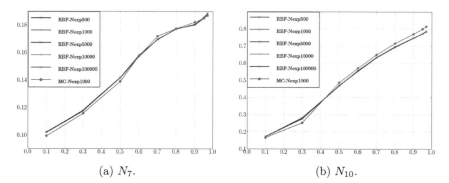

(a) $N_7$.                                       (b) $N_{10}$.

Figure 7.17: Approximated probability distribution function of the criterion EPS in the test nodes $N_7$ and $N_{10}$ using `pan-eps-small` compared with the results of a quasi-MC simulation using 1000 sampling points.

quantiles based on the metamodel converge quickly. Hence, a small sampling size is already sufficient. Moreover, the distribution function approximated with the metamodel is in good accordance with the one approximated by a quasi-MC simulation.

Figure 7.18 gives a detailed view on the results in mesh node $N_4$. It compares the metamodel based results estimated with Nexp = 100000 samples with the quasi-MC solution. Using the model `pan-eps-big` representing more parameters and more sampling points, the distribution function approaches the Monte Carlo solution. We observe only a slight difference in the second decimal place between the approximated distribution functions shown.

Figure 7.19 compares the approximated 0.9—quantile estimated by a quasi-MC simulation with the solution estimated by the $P^2$-algorithm based on the model `pan-eps-small` on the entire geometry. This shows that the solution using `pan-eps-small` is already sufficient. Using the model `pan-eps-big` shows almost no differences to the quasi-MC solution.

The comparison of the estimated quantiles based on a quasi-MC simulation with the ones estimated based on `pan-thickness-small` show almost no differences. Thus, the results are not shown here, because a subsequent optimization is not part of this work.

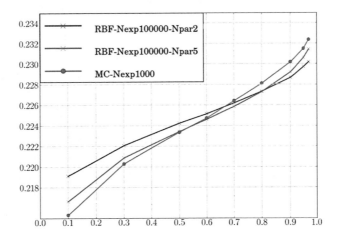

Figure 7.18: Comparison of the approximated probability distribution function in test node $N_4$ based on `pan-eps-small`, `pan-eps-big`, and a quasi-MC simulation.

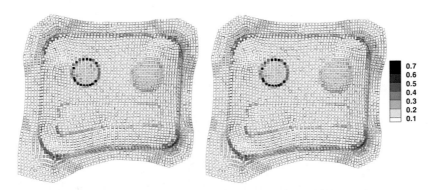

Figure 7.19: Comparison of the 0.9−quantile estimated by a quasi-MC simulation (left) with the solution estimated by the $P^2$-algorithm based on `pan-eps-small` (right).

## 7.2.3   Conclusions

We have observed that the small forecast model `pan-eps-small` based on two parameters, which predicts the criterion EPS, already gives a valuable insight into the process behavior. The local correlations between parameters and test nodes are represented well. This confirms the effectiveness of the parameter classification procedure proposed. Increasing the number of sampling points as well as taking more parameters into account leads to a significant improvement of the overall prediction results, as we have seen using `pan-eps-big`. Increasing the number of sampling points further may increase the forecast quality even more. Additionally, we have seen that the average accuracy obtained is in good agreement with the derived estimator of the proportion of the overall variation explained (cf. Section 4.1.3). Specifically, local regions with higher errors have been identified also with the local total sensitivities measured by the parameter classification procedure developed.

The three models investigated for the fundamental forming example illustrate clearly the benefits of the parameter classification and dimension reduction methods. Moreover, they show that the forecast models used are able to predict the criteria with user controlled accuracy.

Furthermore, we have demonstrated that the probability distribution function can be approximated efficiently. Using the high-quality metamodels derived, the distribution function approaches the quasi-MC solution.

| Method | Nexp | T(Simulations) | T(Statistics) |
|---|---|---|---|
| Quasi-MC | 1000 | 36.1 days (sequential)<br>3.6 days (10 runs in parallel) | 2.39sec |
| PRO-CHAIN | 66 | 2.38 days (sequential)<br>5.72 hours (10 runs in parallel) | 1.03sec |

Table 7.5: Comparison of the computational effort to approximate the probability distribution function of the criterion EPS between a quasi-MC method and PRO-CHAIN using `pan-eps-big`. $T(\cdot)$ denotes the computational time.

Finally, we compare the computational effort of both methods to approximate the distribution function by means of nine $q$-quantiles, with $q \in [0.1, 0.97]$. As already stated, the quasi-MC procedure uses the Halton sequence with Npred $= 1000$ samples. The approximation with the PRO-CHAIN methodology is exemplified with the model `pan-eps-big`.

A single simulation run of this forming example takes approximately 52 minutes performed on a Linux 2.6 GHz HPC Cluster using one compute node. Thus, the required number of simulation runs determines the overall computational time for both methods.

In the case of the PRO-CHAIN methodology, the time for computation of statistics includes the time for the setup and the evaluation of the metamodel with Npred = 1000 samples according to Algorithm 6.1.

The comparison of the computational time demonstrates that an enormous speed-up of 15 is achieved using the new methodology developed to compute full statistical information of the solution compared with a state-of-the-art quasi-MC solution.

## 7.3   Process Chain Forming-to-Crash

As already stated, there is a high industrial need for efficient and thorough analysis methods to identify the impact of parameter variations on the relevant criteria in entire process chains in order to achieve a robust design. Results obtained by these analysis and prediction methods are particularly important for parts of a car with a potentially critical influence in crashes. One of these parts is the B-pillar, which consists of several formed and connected metal blanks [24].

We investigate in detail a forming-to-crash process chain of a B-pillar as a realistic example from the automotive industry. This process chain represents a metal forming scenario as first processing step and a crash scenario as second (and final) processing step. The PRO-CHAIN methodology, with all of its individual components proposed, has been tested successfully on this process chain.

A general overview of the application considered, especially, the parameters involved, the material model, and the damage model is given in Section 7.3.1. The results obtained by the methodology developed are shown in the subsequent sections. The forming processing step is analyzed in Sections 7.3.2 to 7.3.3. The subsequent crash processing step is analyzed in Sections 7.3.4 to 7.3.5. In particular, the results of a forecast model taking the forming history into account are presented. We summarize the main results and benefits of the new methodology for the process chain in Section 7.3.6.

### 7.3.1   ZStE340 Metal Blank of a B-Pillar

We apply the PRO-CHAIN methodology newly developed to a two step process chain of a metal blank formed to a part of a B-pillar, which has been provided by Daimler AG. The first processing step is a deep drawing process, cf. Section 2.3. As second processing step a tension-bearing test of the B-pillar under crash loading was performed, referred to as crash processing step in the following.

**Remark 7.8.** *In this component test, the B-pillar has been overloaded absolutely unrealistically until a failure occurs. This is performed in order to analyze the material and damage behavior in detail, and in order to demonstrate that the new methodology is able to predict possible failure correctly.*

First results of individual aspects of the PRO-CHAIN methodology developed and applied to this example have been published in the conference contributions and papers [47, 24, 23, 25, 111, 112, 113]. In [47, 24, 23, 25, 111], predominantly, results of the parameter classification procedure are presented and only single simulation results have been mapped. The papers [112, 113] present first results on forecast models for single processing steps.

The material considered is a micro-alloyed fine-grain steel (ZStE340), which is widely used in the industry because of its formability [47, 6]. For details about the material characterization refer to [47, 25].

The forming and the crash simulations have been carried out by means of the FE simulation code LS-DYNA®. The simulation setups of both steps are illustrated in Figure 7.20.

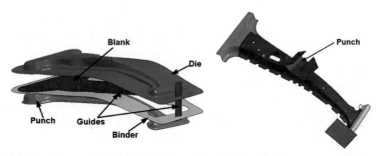

(a) Forming tools and blank before the forming process.

(b) Simulation setup of the tension-bearing test.

Figure 7.20: Simulation setup of the process chain forming-to-crash.

In Figure 7.21 the part of the forming mesh relevant for the mapping to the crash mesh is highlighted in blue. The forecast model should be highly accurate in this domain. Note that the remaining part of the mesh is cut off, so that the predicted distributions in this domain do not contribute to the input for the crash simulations.

The results of the forming simulation run are mapped to the crash mesh as described in Section 2.2.3 so that they serve as input distribution for

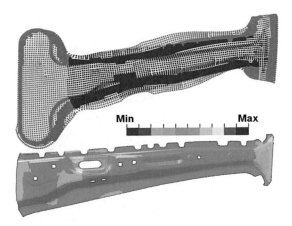

Figure 7.21: Part of the forming mesh (blue, top) relevant for the mapping to the crash mesh together with mapped thickness serving as input for a crash simulation (bottom).

crash simulations. The mapped distribution of thickness resulting from the nominal forming simulation run is exemplified in Figure 7.21.

The accuracy of simulation results depends strongly on the material and damage model used. To consider anisotropic material behavior, the three parameter material model from Barlat is applied, details can be found in [25] and references therein.

Furthermore, we have to use a single damage model which is suitable for both, forming and crash simulation. Biaxial loading is dominant in deep drawing processes, whereas uniaxial loading mainly appears in crash processes [47]. For this reason, the damage model should be able to reflect the influences by different loading conditions. Therefore, we use the *bi-failure damage model*, which is based on a modified Johnson-Cook model. It was developed by Fraunhofer IWM, especially for this coupling task, see [47, 118] and references therein. We investigate the influences due to parameter variations of this numerical damage model explicitly in Section 7.3.3. In the bi-failure model, the fracture strain $\epsilon_f$ is defined as a function of the stress triaxiality. The domain is divided into two regions, dependent on a triaxiality $T_{\text{trans}}$, for dimple rupture at high triaxialities and shear failure at low triaxialities, that is,

$$\epsilon_f = \begin{cases} d_1 + d_2 \exp(-d_3 T), & T > T_{\text{trans}} \\ d_{\text{Shear1}} + d_{\text{Shear2}}|T|^{m_2} + d_{\text{Shear3}}(-T)_+^{m_3}, & T < T_{\text{trans}}, \end{cases} \tag{7.6}$$

with $(x)_+ = x$ if $x$ is positive and zero otherwise. The fracture strain at $T = T_{\text{trans}}$ has to be specified. Failure occurs, if the criterion damage, computed using this failure model, reaches the critical value of 1, for details refer to [47]. In total, the bi-failure damage model involves four indepen-

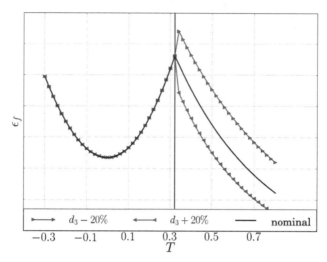

Figure 7.22: Influences of variations of the parameter $d_3$ on the bi-failure damage model. The vertical line corresponds to $T = T_{\text{trans}}$.

dent parameters $(d_1, d_2, d_3, d_{\text{Shear1}})$ which can be subject to scatter and are included in the sensitivity analysis. The variables $d_{\text{Shear2}}$, $d_{\text{Shear3}}$, $m_2$, $m_3$ are fixed or calculated automatically to ensure continuity of the strain function. The strain function is not differentiable in the point $T = T_{\text{trans}}$, which is obvious from Figure 7.22. This leads to an instability, that is, small variations of $T_{\text{trans}}$ have a high impact on the resulting fracture strain, and, thus, on the criterion damage. We have confirmed this instable behavior with numerical experiments. Thus, we assume the parameter $T_{\text{trans}}$ of the numerical damage model to be fixed and not subject to variations, since the new methodology assumes a stable process behavior.

An experimental validation of the material and damage model was performed by means of a component test [47]. To achieve a combination of bending with superimposed tension, the B-pillar was supported at both ends by revolvable bearings, see Figure 7.23. The load application occurred path controlled.

In the following analysis of the process chain forming-to-crash with the PRO-CHAIN methodology developed, 14 material and process parameters have been varied as listed in Table 7.6. The range of variation for each parameter has been determined partly by experiments and partly due to literature so that it reflects variations arising in practice. The metal blank has a nominal sheet thickness of 1.75 mm. As results on the entire mesh, we evaluate the distributions of thickness, effective plastic strain (EPS), and

Figure 7.23: Experimental setup of the tension-bearing test (from [47]).

| Parameter type | | Parameter | Variation range |
|---|---|---|---|
| Material | Sheet thickness | $t$ | $\pm 10\%$ |
| | Hardening | $K, n, \varepsilon_0$ | $\pm 10\%$ |
| | Anisotropy | $R_0, R_{45}, R_{90}$ | $\pm 10\%$ |
| Damage | Damage | $d_1, d_2, d_3, d_{\text{Shear1}}$ | $\pm 20\%$ |
| Process | Friction | $\mu$ | $\pm 50\%$ |
| | Hold down force | $F^H$ | $\pm 10\%$ |
| | Drawing force | $F^D$ | $\pm 10\%$ |

Table 7.6: Parameters with their range of variation in the process chain forming-to-crash.

damage. A typical scalar criterion, namely the maximal force, is considered additionally in the crash step.

**Remark 7.9.** *In the forming processing step adaptive mesh refinement is used. Hence, we specify the mesh of the nominal simulation run as reference mesh in the construction of the database* **M** *in the following analysis, cf. Section 2.2.3.*

In the forming process, we have observed that the parameter sheet thickness ($t$) has a very strong influence on all criteria considered. Specifically, the variation of the parameter thickness itself already accounts for 68 percent of the overall variation due to parameter variations with respect to the criterion thickness. Thus, a simple metamodel including only the variation of $t$ will already lead to good predictions of the criterion thickness. In order to analyze in detail the effects of the other parameters involved, we assume that the sheet thickness is constant and not a varying parameter in the following analysis. Hence, 10 material and damage parameters, and 3 process parameters are involved in the analysis of the forming processing step. In the crash process, all material and damage parameters are considered as random parameters. That is, the influence of the parameter sheet thickness is investigated in the sensitivity analysis of this processing step, too.

**Remark 7.10.** *In the following, all resulting distributions of the criteria considered are scaled to* $[0, 1]$ *for data confidentiality reasons. Furthermore, in the crash processing step, local resulting distributions are shown on the initial, not deformed mesh, unless stated otherwise.*

## 7.3.2 Parameter Classification of the Forming Step

We apply the parameter classification procedure developed in Section 4 to the forming processing step. The accumulated sensitivity measures $S1^j$ and $S2^j$ for the three criteria considered, EPS, thickness, and damage, show similar results, which can be seen from Figure 7.24. Specifically, the order of the parameters for EPS and thickness is almost the same. Considering the criterion damage, three parameters of the damage model, $d_1, d_2$ and $d_3$, have strong influences, as expected. The order of the other parameters remains the same.

The complete classification results for the three criteria considered are shown in Table 7.7 and Table 7.8. The linear and total importance classes are identical for the criteria EPS and thickness.

The assignment of the parameters to linear and total importance classes differs considering the criterion damage. The parameters in the first nonlinearity class $\mathrm{NLClass}_1 = \{R_0, R_{45}, R_{90}, \varepsilon_0, n, \mu, F^H, F^D\}$ influence the criterion damage in a nonlinear way. The remaining parameters, specifically, the damage parameters, are assigned to the second nonlinearity class $\mathrm{NLClass}_2$, i.e., they influence the criterion damage linearly on average. The nonlinearity classes of the criteria EPS and thickness differ slightly, most of the parameters are assigned to the nonlinearity class $\mathrm{NLClass}_1$. Considering the criterion EPS, we obtain $\mathrm{NLClass}_2 = \{K\}$, and considering the criterion thickness, we obtain $\mathrm{NLClass}_2 = \{K, \mu\}$, these parameters behave linear on average.

Since the forming process affects the criteria very local, it is essential to look at the local influences as well. As described, this is supported by the parameter classification developed, and demonstrated in the remaining of this section. The local distribution of the nonlinearity measure, given by Equation (4.18), of the strongly influential parameter friction ($\mu$) with respect to the criterion EPS is illustrated in Figure 7.25. The nonlinear influences, indicated with negative values, are dominating the linear ones in local regions of the B-pillar. Specifically, the nonlinear influences are strong in regions where the blank is highly deep drawn, and/or high strains occur, for example, in the regions marked with A, B, C and D. This strong nonlinearities are identified also with the accumulated nonlinearity measure, since $\mu$ is assigned to the first nonlinearity class.

If only the variations of the parameters of the first total importance class are taken into account in a forecast model for criterion EPS or thickness,

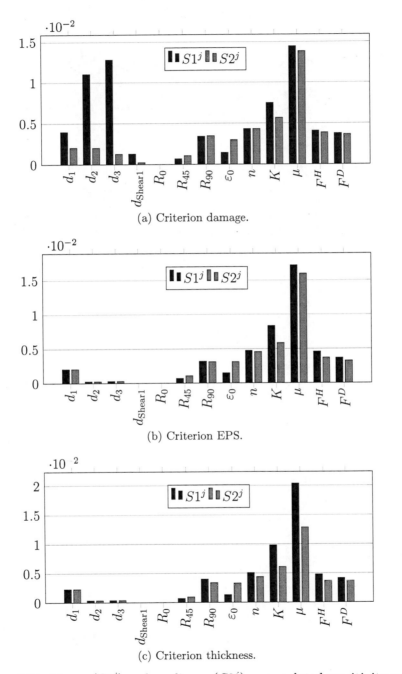

(a) Criterion damage.

(b) Criterion EPS.

(c) Criterion thickness.

Figure 7.24: Linear $(S1^j)$ and nonlinear $(S2^j)$ accumulated sensitivity measures for the three criteria considered in the forming processing step.

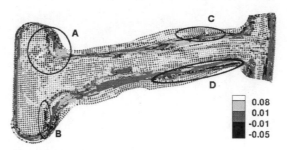

Figure 7.25: Local nonlinearity measure of parameter friction with respect to the criterion EPS. The nonlinear influences, indicated with negative values, are dominating the linear ones in local regions of the B-pillar, for example, marked with A, B, C and D.

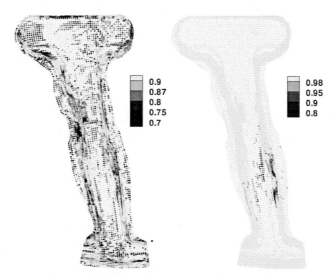

Figure 7.26: Proportion of the overall variation explained by the union of the parameters in the first total importance class (left), and the first and second total importance class (right) considering the criterion EPS.

| Class | Parameter | EPS % | Thickness % |
|---|---|---|---|
| LIClass$_1$ | $R_{90}$, $n$, $K$, $\mu$, $F^H$, $F^D$ | 89.7 | 90.04 |
| LIClass$_2$ | $d_1$, $R_{45}$, $\varepsilon_0$ | 98.8 | 98.55 |
| LIClass$_3$ | $d_2$, $d_3$ | 100 | 100 |
| LIClass$_4$ | $d_{\text{Shear1}}$, $R_0$ | 100 | 100 |
| TIClass$_1$ | $R_{90}$, $n$, $K$, $\mu$, $F^H$, $F^D$ | 87.04 | 86.88 |
| TIClass$_2$ | $d_1$, $R_{45}$, $\varepsilon_0$ | 98.72 | 98.37 |
| TIClass$_3$ | $d_2$, $d_3$ | 100 | 100 |
| TIClass$_4$ | $d_{\text{Shear1}}$, $R_0$ | 100 | 100 |

Table 7.7: Forming-to-crash: Parameter classification results for the criteria EPS and thickness. % denotes the average proportion of the overall variation explained by the union of the parameters up to the class considered.

| Class | Parameter | Damage % |
|---|---|---|
| LIClass$_1$ | $d_2$, $d_3$, $K$, $\mu$ | 65.92 |
| LIClass$_2$ | $d_1$, $R_{90}$, $n$, $F^H$, $F^D$ | 94.04 |
| LIClass$_3$ | $d_{\text{Shear1}}$, $\varepsilon_0$, $R_{45}$ | 100 |
| LIClass$_4$ | $R_0$ | 100 |
| TIClass$_1$ | $d_2$, $d_3$, $n$, $K$, $\mu$, $F^H$, $F^D$ | 81.21 |
| TIClass$_2$ | $d_1$, $R_{90}$, $\varepsilon_0$ | 96.39 |
| TIClass$_3$ | $d_{\text{Shear1}}$, $R_{45}$ | 100 |
| TIClass$_4$ | $R_0$ | 100 |

Table 7.8: Forming-to-crash: Parameter classification results for the criterion damage. % denotes the average proportion of the overall variation explained by the union of the parameters up to the class considered.

more than 85 percent of the overall variation is expected to be explained on average. Adding the parameters of the second total importance class, more than 98 percent of the overall variation is expected to be explained on average. The percentages of the overall variation which are expected to be explained locally by the parameters of the first, and the first two total importance classes, are illustrated in Figure 7.26. Taking only the parameters of the first total importance class into account, the actual variation explained is much lower than the expected average variation explained in some regions of the B-pillar. Whereas, only in a small region of the B-pillar, the proportion of variations explained is below the expected average variation explained (98 percent), taking the parameters of the first two total importance classes into account. This region is partly cut off in the initial mesh of the crash processing step.

Next, we investigate the local influences on the criterion damage in detail. The local distribution of the sensitivity results considering the influ-

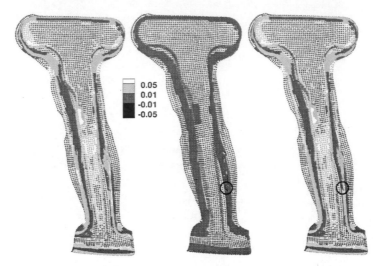

Figure 7.27: Influences of variations of parameter $d_3$ on the criterion damage. Left: linear influences; center: nonlinear influences, which are almost zero except in the region highlighted; right: difference between linear and nonlinear influences.

ences due to variations of the damage parameter $d_3$ on the criterion damage are presented in Figure 7.27. The sensitivity matrix $S1$ (left) is compared with the nonlinear influences $S2$ (center), resulting in the local nonlinearity measure (right). The nonlinear influences are high only in a small region causing a nonlinear behavior of $d_3$ only in this region highlighted with a black circle. The assignment of $d_3$ to the second nonlinearity class accords well with the local results obtained.

The damage parameters $d_2$ and $d_3$ are assigned to the first total importance class. Hence, they influence the criterion damage strongly (cf. Table 7.8). These parameters belong to the numerical bi-failure damage model used in both processing steps (cf. Equation (7.6) and Figure 7.22). Thus, we investigate this damage model numerically, in detail in Section 7.3.3. For this purpose, we assume that only the four damage parameters involved are subject to scatter. All other parameters are assumed to be constant. Under this assumption, the proportion of the overall variation in the criterion damage which can be explained taking the influences due to parameters $d_2$ and $d_3$ into account is 85 percent on average. Figure 7.28 shows the proportion of variation explained, determined by the sum of total sensitivities

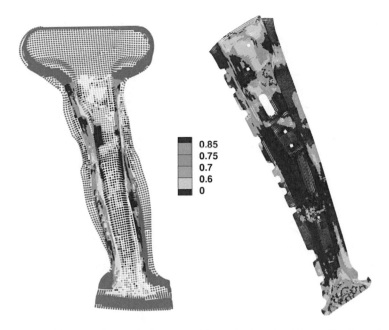

0.85
0.75
0.7
0.6
0

Figure 7.28: Proportion of the overall variation in the criterion damage which is expected to be explained by the sum of the parameters $d_2$ and $d_3$ considering the bi-failure damage model. Left: forming processing step; right: crash processing step taking the forming history into account.

(Equation (4.27)), analyzing the forming processing step (left) and the crash processing step taking the forming history into account (right). Specifically, the majority of the part relevant for mapping to the crash mesh is represented well taking variations due to the parameters $d_2$ and $d_3$ into account. Some parts at both ends of the B-pillar are not expected to be explained by the two parameters. However, we show that the forecast models derived in the following section give appropriate results also in these regions.

### 7.3.3  Forecast Models

We require fast and accurate forecast models for the final distributions of the three criteria EPS, thickness, and damage in order to enable a propagation of all relevant scatter information to the next processing step.

For this purpose, we setup several forecast models, based on the results of the parameter classification procedure and the subsequent processing of the database (cf. Chapter 4 and Chapter 5). In detail, we investigate the four models listed in Table 7.9. Since the first total importance classes for the criteria EPS and thickness are identical, we can use the same ensem-

| Model name | Criterion | Parameters | % | Nexp |
|------------|-----------|------------|-----|------|
| B-pillar-eps-small | EPS | $R_{90}, n, K, \mu, F^H, F^D$ | 87.04 | 58 |
| B-pillar-eps-big | EPS | $R_{90}, n, K, \mu, F^H, F^D$ | 87.04 | 87 |
| B-pillar-thickness | thickness | $R_{90}, n, K, \mu, F^H, F^D$ | 86.88 | 87 |
| B-pillar-damage | damage | $d_2, d_3$ | 85* | 20 |

Table 7.9: Overview of different metamodels investigated for the forming processing step. % refers to the average percent of variation which is expected to be explained by each model. Nexp denotes the number of samples used to create the corresponding metamodel. *In this scenario, only the four damage parameters are assumed to be random.

ble of simulation results to predict both criteria. Therefore, a metamodel which takes the Npar = 6 parameters of the first total importance class into account is set up. Recall that almost all parameters are assigned to the first nonlinearity class, that is they behave nonlinear on average. Thus, according to the iterative extension of the database, developed in Section 5.2, we start with $C = 2$ leading to Nexp = 58 samples. Since this model B-pillar-eps-small does not give sufficient results for the prediction of the criterion EPS, we iterate the processing of the database again. Finally, we use $C = 3$ and Nexp = 87 samples leading to the models B-pillar-eps-big and B-pillar-thickness to predict the criteria EPS and thickness. Both models are expected to explain on average at least 86 percent of the overall variation in the criterion considered.

As already stated, we investigate in detail the bi-failure damage model numerically. For ths purpose, only the four damage parameters are varied, all other parameters are assumed to be constant. Therefore, we consider the model B-pillar-damage, including the parameters $d_2$ and $d_3$. This model is based on the processed database containing a minimal number of Nexp = 20 samples according to Section 5.2. The model is expected to explain at least 85 percent of the variation on average.

The accelerated RBF metamodel is used in all cases as forecast model, as introduced in Section 6.1. In order to analyze the local process behavior and to measure the prediction error locally, we specify nine mesh nodes as test nodes, highlighted in Figure 7.29. These mesh nodes are selected at interesting parts of the B-pillar. In particular, some of the test nodes may become critical in forming and/or crash. In this case, EPS values may become very high, or the component may be thinned severely, locally during the forming process. Nodes in the border regions of the B-pillar are not considered, since these do not contribute to the mapping to the crash mesh.

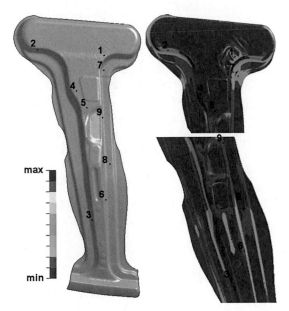

Figure 7.29: Nine selected test nodes on the formed B-pillar (left) and detailed view on the resulting distribution of the criterion EPS (right).

**Prediction of the Criterion Effective Plastic Strain**

We consider the models B-pillar-eps-small and B-pillar-eps-big to predict the criterion EPS. Both models take the six parameters of the first total importance class into account. Hence, they are expected to explain at least 87 percent of the overall variation on average.

Additionally, we compute the estimator of the average prediction error in an arbitrary point according to Corollary 6.14 (cf. Section 6.4) exemplary using the model B-pillar-eps-big. As already stated, the parameter space reduction is based on $\varepsilon_\sigma = 0.13$. Furthermore, we set the maximal acceptable loss of information due to the SVD of the database to two percent, i.e., $\varepsilon_{\mathrm{SVD}} = 0.02$. The averaged total sensitivities of all parameters is given by $\mathrm{var_g} = \sum_{j=1}^{\mathrm{Npar}} \mathrm{TI}^j = 0.09$, which is bigger than the average variation in $\mathbf{M}_i$, cf. Theorem 6.12. Note that the maximal variation is much higher in local regions. Finally, the average prediction error of the RBF metamodel B-pillar-eps-big is measured. Hence, the model is evaluated in Npred = 100000 random points, and the average model tolerance over all points is computed, which leads to $\mathrm{tol}(\widetilde{\mathbf{g}}) = 0.0038$. In total, the average prediction error is derived as

$$\frac{1}{\text{Nnodes}} \| \mathbf{g}(\mathbf{P}) - \widetilde{\mathbf{g}}(\mathbf{P}) \|_1$$
$$\leq (1 - (1 - \varepsilon_\sigma)(1 - \varepsilon_{\text{SVD}})) \, \text{var}_\mathbf{g} + \text{tol}(\mathbf{P})$$
$$= (1 - (1 - 0.13)(1 - 0.02)) \, 0.09 + 0.0038 = 0.017. \qquad (7.7)$$

If the acceptable error due to the SVD is increased to 12 percent, corresponding to $k = 26$ singular values retained in the model B-pillar-eps-big, the estimated average error will increase to 0.025. The following analysis shows that both error thresholds are fulfilled, and that the actual average error is much lower. Indeed, the actual error is lower than the derived upper threshold in 95 percent of the mesh nodes in arbitrary points.

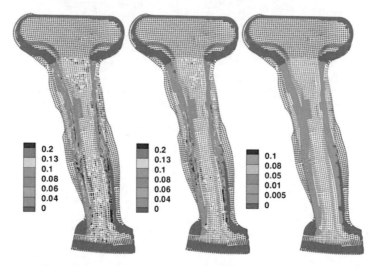

Figure 7.30: Comparison between the relative prediction errors in test point $\mathbf{P}_{13}$ using B-pillar-eps-small (left) and the relative prediction errors (center) together with the corresponding absolute prediction errors (right) in this point using B-pillar-eps-big.

In order to validate the accuracy of the forecast models, we specify 14 randomly selected out-of-sample points, and perform the corresponding simulation runs. We observe that Nexp = 58 samples are not enough to reach the expected 87 percent threshold comparing the predicted results with the simulation results for the test points specified. The model B-pillar-eps-big based on the extended database containing Nexp = 87 samples gives much better results. For example, a comparison between the relative prediction errors in test point $\mathbf{P}_{13}$ using model B-pillar-eps-small and B-pillar-eps-big is illustrated in Figure 7.30. The relative prediction errors are smaller, when the model with more samples is used. For comparison, also the absolute prediction errors in this test point using the model `B-pillar-eps-big` are shown. This demonstrates that the still higher relative errors correspond to rather low absolute errors, since the EPS value itself is low in these regions. Thus, the big model gives sufficient prediction results.

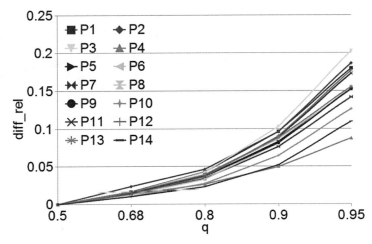

Figure 7.31: Approximated distribution function of relative prediction errors in the 14 test points $\mathbf{P}_1, \ldots, \mathbf{P}_{14}$ using `B-pillar-eps-big`.

The distribution of the local relative prediction differences $\mathbf{diff}_{rel}$, according to Equation (6.2), in the 14 test points specified is shown in Figure 7.31. The average relative differences, i.e., $q = 0.5$, are almost zero. Moreover, 90 percent of the local relative differences are below ten percent for all test points considered. Few higher relative errors remain. However, the absolute differences, corresponding to higher relative errors, are below 0.02, that is, rather small. This can also be seen in Figure 7.30 for the test point $\mathbf{P}_{13}$. In particular, 95 percent of the absolute errors are below the estimated average error (Equation (7.7)). Especially, the estimated threshold of the average error is fulfilled. In summary, the expected average amount of total variation due to parameter variations explained (87 percent) is reached. Yet, more than 90 percent of the relative errors are below this threshold. Adding more samples to the construction of the model is not expected to give even better results, since the local areas with higher errors cannot be fully explained by the six parameters taken into account. Thus, we evaluate the model `B-pillar-eps-big` further in the remaining of this paragraph.

The parameter classification has revealed that the friction coefficient is the most influencing parameter on the criterion EPS, and that the influ-

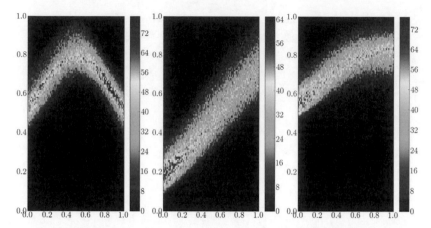

Figure 7.32: Correlation between parameter friction (x-axis) and criterion EPS (y-axis) in the test nodes $N_1, N_6, N_8$ (from left to right). The color coding corresponds to the frequency of occurrence of a certain EPS value. Both, the values of friction and resulting EPS are scaled to $[0, 1]$.

ences are nonlinear locally (cf. Section 7.3.2). Hence, we investigate the correlation between the parameter friction and the criterion EPS in detail by evaluating the model `B-pillar-eps-big` with a Monte Carlo procedure. We observe that the friction coefficient influences the EPS values locally in different ways. For example, Figure 7.32 shows that the friction coefficient has a monotonous increasing influence in test node $N_6$. Thus, increasing the friction coefficients causes an increase of the EPS values. Whereas, in test node $N_8$ the behavior is slightly non-monotonous, and in $N_1$ strongly nonlinear. Specifically, two clusters are obvious, one at small EPS values, and one at high EPS values (red). That is, a large part of the samples lead to rather small values, and another large part of the samples lead to rather high EPS values in this local region. Hence, it is important to analyze the local behavior represented by the local probability distribution function in order to improve the robustness of the component.

The previous results have been obtained with the full number of singular values in the SVD of the corresponding database using `B-pillar-eps-big`. In order to benefit from the acceleration of the RBF metamodel, we analyze the prediction error of the SVD using only few singular values. The average prediction error $\text{err}_2(k)$, derived in Section 5.3, is graphed in Figure 7.33 as a function of the number of singular values retained. The prediction error decreases fast at the beginning of the function until $k = 5$. Then, the curve decreases smoothly against zero.

The first and the fifth mode weighted with the corresponding singular value, i.e., $\lambda_1 \mathbf{U_1}$ and $\lambda_5 \mathbf{U_5}$, correspondingly, are illustrated in Figure 7.34.

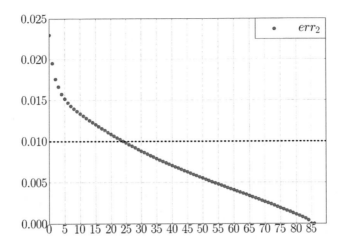

Figure 7.33: Average absolute approximation error $\text{err}_2(k)$ of the SVD of the database containing Nexp = 87 samples as a function of the number of singular values retained.

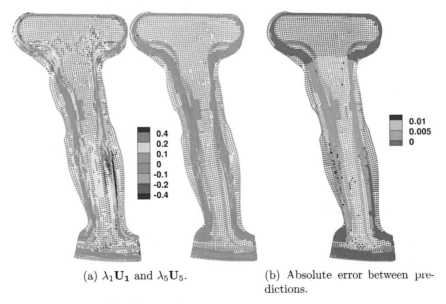

(a) $\lambda_1 \mathbf{U_1}$ and $\lambda_5 \mathbf{U_5}$.

(b) Absolute error between predictions.

Figure 7.34: 1-st and 5-th mode of a singular value decomposition of the database of model B-pillar-eps-big weighted with corresponding singular values. Additionally, the absolute differences between a prediction using the full database and the compressed database is shown.

Recall that the average result in the database has been subtracted before the SVD is applied. Hence, the first mode is already a good representative of the deviations from the average result. The fifth mode contributes only to a very small region of the B-pillar.

If we determine the absolute acceptable loss of information by $\varepsilon_{\text{SVD,abs}} = 0.01$, highlighted in 7.33, it will be sufficient to retain $k = 26$ singular values. All remaining singular values are omitted, i.e., set to zero. An evaluation of the forecast model using this compressed database instead of the full database should lead to, at least, an average increase of the absolute prediction difference by $\varepsilon_{\text{SVD,abs}} = 0.01$. We compare the predicted result using the full database with the predicted result using the compressed database. The absolute difference between these two predictions is exemplified in Figure 7.34 on the right-hand side for the test point $\mathbf{P}_{13}$. As expected, the differences are below the threshold in the majority of the nodes. Hence, using the compressed database with $k = 26$ singular values instead of the full database, an appropriate accuracy is retained. The estimated threshold of the average error according to Equation (7.7) is fulfilled. Therewith, a compression factor of $C_R = \frac{87}{27} = 3.22$ is achieved. That is, only 30 percent of the memory needed to store the full database is required. Additionally, a speed-up of 3.22 can be achieved, when the metamodel using the compressed database is evaluated numerously. This is the case, when statistics of the criterion are approximated by means of the forecast model.

The model `B-pillar-eps-big` is used to approximate the probability distribution function of the criterion EPS. The approximated probability distribution functions in the test nodes $N_1, N_6$, and $N_8$ are given in Figure 7.35.

**Remark 7.11.** *In order to approximate the probability distribution function of the criteria considered, the corresponding metamodel is evaluated by means of Algorithm 6.1 with* Npred $= 100000$ *randomly distributed sampling points, and the q-quantiles with $q \in [0.05, 0.95]$ are estimated.*

This illustrates that the probability distribution is locally different. To be more specific, the EPS value is low in the test nodes $N_6$ and $N_8$, whereas the EPS value is higher in test node $N_1$. Moreover, there is almost no variation in test node $N_8$, but a high variation of EPS values in test node $N_1$, comparing the 0.1-quantile and the 0.9-quantile. Hence, this local area is strongly influenced by the forming process. This area has also been identified by the parameter classification procedure as strongly nonlinear influenced by the friction coefficient, cf. Figure 7.25.

This is confirmed, when the probability distribution on the entire mesh is investigated. For example, the resulting median, the 0.1-quantile, and the 0.9-quantile are illustrated in Figure 7.36. The 0.9-quantile shows high EPS

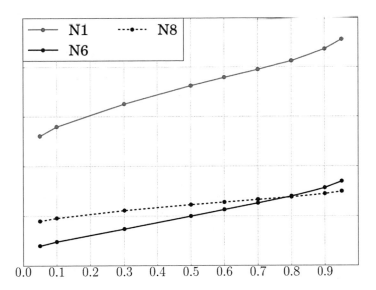

Figure 7.35: Approximated probability distribution function of the criterion EPS in the test nodes $N_1$, $N_6$, and $N_8$.

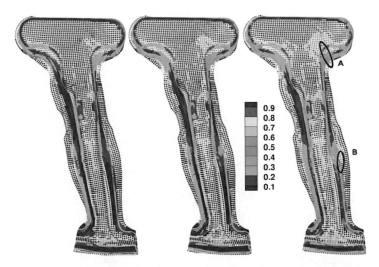

Figure 7.36: 0.1-quantile (left), 0.5-quantile, (center) and 0.9-quantile (right) of the distribution of the criterion EPS using B-pillar-eps-big. The median is scaled to $[0, 1]$, the other quantiles are scaled relative to the maximal value of the median.

values around test node $N_1$, that is in ten percent of the realizations the local EPS values in this region are high. Note that the distribution values are scaled to $[0, 1]$. Furthermore higher EPS values occur with ten percent probability in the areas marked by A and B. However, the higher values in the region B do not affect the behavior in a subsequent crash analysis, because this region is cut off in the crash mesh.

### Prediction of the Criterion Thickness

The same ensemble of simulation runs can be used in order to predict the criterion EPS and thickness, since the same parameters are assigned to the first total importance class in both cases. We investigate the model **B-pillar-thickness** to predict the criterion thickness. The 14 specified test points are used again to validate the precision of the forecast model.

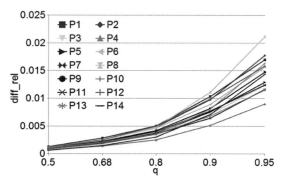

(a) Approximated distribution of relative prediction errors.

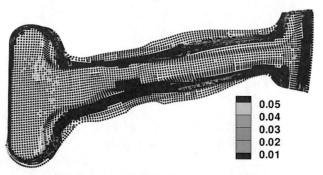

(b) Relative errors in the test point $\mathbf{P}_{13}$.

Figure 7.37: Approximated distribution function of relative prediction errors for the 14 test points considered together with local relative prediction errors in the test point $\mathbf{P}_{13}$.

The parameters of the first total importance class represent the overall resulting variation in the criterion thickness very well, except for a few local regions. This can be seen from Figure 7.37. It shows the approximated distribution of the relative prediction differences in all test points, which is below 2.5 percent with 95 percent probability. The few higher errors arise in local regions around test nodes $N_1$ and $N_2$. Additionally, few higher errors arise locally at both sides of the B-pillar. These do not contribute to the input distribution in the subsequent crash processing step, since these regions are cut off.

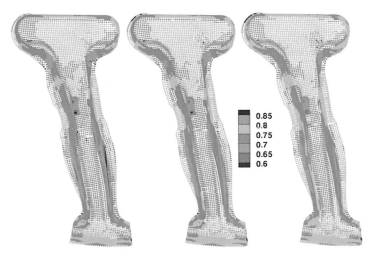

Figure 7.38: 0.1-quantile (left), 0.5-quantile (center), and 0.9-quantile (right) of the distribution of the thickness using **B-pillar-thickness**. The median is scaled to $[0, 1]$, the other quantiles are scaled relative to the maximal value of the median.

In the following, the probability distribution of the criterion thickness is approximated. For this purpose, the model **B-pillar-thickness** is evaluated (cf. Remark 7.11). The approximated median, the 0.1-quantile, and the 0.9-quantile of the probability distribution of the criterion thickness are presented in Figure 7.38. In detail, local regions of the B-pillar can be identified, where possible defects may occur (cf. Section 2.3). For example, a local thinning at the right side of the B-pillar can be observed with a probability of ten percent, that is the probability that the thickness is below the given value in the 0.1-quantile (scaled to $[0, 1]$) is ten percent. Furthermore, the 0.9-quantile shows that a ten percent probability exists that the local thickness values around $N_1$ increase further.

**Prediction of the Criterion Damage**

As already stated, the damage parameters $d_2$ and $d_3$ influence the criterion damage strongly. Therefore, we investigate the bi-failure damage model numerically, in detail in the following scenario. For this purpose, only the four damage parameters are varied, all other parameters are assumed to be constant.

Thus, we consider the model `B-pillar-damage-small` including the parameters $d_2$ and $d_3$. This model is based on the processed database containing a minimal number of samples Nexp = 20 according to Section 5.2. The model is expected to explain at least 85 percent of the variation on average, the local distribution of expected proportion of variation explained has been shown in Figure 7.28.

We specify five randomly selected out-of-sample points $\mathbf{P}_1, \ldots, \mathbf{P}_5$ within the entire parameter space as test points in order to validate the predicted results. These test points include the full uncertainty caused by the four damage parameters considered in this model.

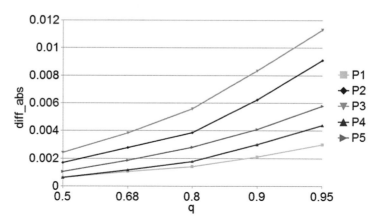

Figure 7.39: Approximated distribution function of absolute prediction errors in the five test points using `B-pillar-damage-small`.

The distribution of the absolute prediction errors $\mathbf{diff}_{\mathrm{abs}}$ in the five test points according to Equation (6.2) is shown in Figure 7.39. The distribution is computed over the entire forming mesh, not only the part relevant for mapping to the crash mesh. The figure shows that the average absolute error is below 0.004, that corresponds to a relative error of four percent in all test points. The 0.95 quantile of the absolute prediction errors is rather small, i.e., below 0.02. Hence, the relative prediction error is at most ten percent with 95 percent probability. This is much better as expected from the average estimator of proportion explained, which is 85 percent.

Moreover, this estimated accuracy is achieved with more than 95 percent probability.

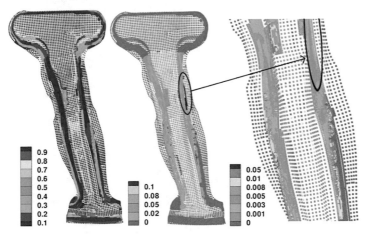

Figure 7.40: Predicted damage distribution (left) and local relative prediction errors (center) in test point $\mathbf{P}_2$ together with detailed view on the corresponding local absolute prediction errors (right) using B-pillar-damage-small.

This high accuracy of the forecast model is also confirmed by the local prediction error, exemplified in Figure 7.40 for test point $\mathbf{P}_2$ on the part of the forming mesh relevant for mapping to crash. The higher relative errors correspond to very low absolute prediction errors, since the criterion damage itself is very low in these regions. The other test points show equal or even better results.

This high-quality metamodel derived is now evaluated in order to compute statistics of the criterion damage. That is, the probability distribution function is approximated by means of $q$-quantiles, cf. Remark 7.11. The median, the 0.1-quantile, and the 0.9-quantile are illustrated in Figure 7.41. Specifically, the figure shows that higher damage values occur very locally. The higher values (green) at both sides of the B-pillar are not relevant for mapping to crash, since these parts are cut off.

Based on the influences of variations of the parameter $d_3$ on the criterion damage, a clustering is performed (cf. Section 4.2). The resulting decomposition into Nclust = 3 clusters is shown in Figure 7.42. Comparing the cluster assignment and the distribution of damage criteria shown in Figure 7.40, a good accordance of cluster $CL_2$ with the regions of high damage values can be seen. Thus, the clustering results can be used in order to reduce the computational effort to compute statistics of the criterion (cf. Section 6.2). To be more specific, Npred = 100000 sampling points

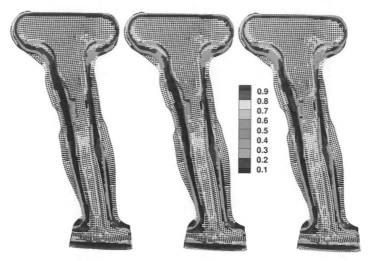

Figure 7.41:  0.1-quantile (left),  0.5-quantile (center),  and 0.9-quantile (right) of the distribution of the criterion damage using B-pillar-damage-small. The median is scaled to $[0, 1]$, the other quantiles are scaled relative to the maximal value of the median.

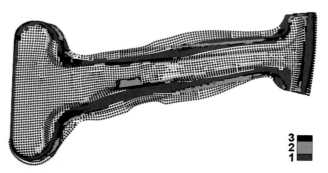

Figure 7.42: Three different clusters obtained by a clustering based on the influences of variations of the parameter $d_3$ on the criterion damage. Cluster one corresponds to almost no influence, clusters two and three to strong influences.

are used only in Nclust = 2 and much fewer sampling points are sufficient in the remaining clusters. This leads to the same results, but reduces the computational effort to estimate the $q$-quantiles considerably.

### 7.3.4 Parameter Classification of the Crash Processing Step

The second processing step is a component test of a B-pillar, in which the B-pillar is overloaded unrealistically until a failure occurs. The analysis of the second processing step starts again with the parameter classification developed. Analogously to the forming processing step, the criteria EPS, damage and thickness are considered.

We want to use a single metamodel for the prediction of a criterion in all timesteps considered, based on a unique parameter classification for all time steps. Therefore, we compare the classification results of several fixed timesteps with the classification results obtained by using the whole database $\widetilde{\mathbf{M}}$ containing all simulation results over time. Therewith, we have observed that the state ts = 90 is a good representative of the parameter influences over time. Thus, we use $\mathbf{M}$ with ts = 90 in the following classification. The failure has already been initiated or progressed in most simulations in this timestep.

**Remark 7.12.** *In order to predict a possible failure properly, the influences due to parameter variations which may lead to this failure have to be taken into account. Moreover, the B-pillar is overloaded absolutely unrealistically in the simulation, so that the final timestep corresponds to a complete crack in the component. Thus, we do not consider the final timestep, because this provides a rather chaotic behavior. In this case, general parameter dependencies are difficult to find, and these do not reflect the overall behavior until the failure occurred. Hence, the forecast models are also constructed to predict the criteria considered until a crack is clearly formed. For this reason, we consider the progress of the process until timestep ts = 90.*

First, we perform the parameter classification with taking the forming history into account. For this purpose, the 2Npar + 1 forming results, precisely, the resulting distributions of the three criteria considered, are mapped to the crash processing step. In this case, we analyze the influences of all 14 parameters. The variations are given by the distributions of the criteria resulting from the forming step.

Then, the parameter classification results of the scenarios, with and without mapping of the forming results, are compared. Therefore, we repeat the parameter classification of the crash processing step without taking the history due to the forming process into account. That is, the crash process is interpreted as single, independent processing step, as it is state-of-the-art. In this case, 11 parameters are varied according to Table 7.6. The three process parameters of the forming step are not involved in the crash step.

The accumulated sensitivity measures $S1^j$ and $S2^j$ for the three criteria considered taking the forming history into account are presented in Figure 7.43. The parameter thickness has the most influence on all criteria considered. As expected, the damage parameters have strong influences only on the criterion damage. Moreover, it can be seen that the friction coefficient, a process parameter of the forming step, has a strong influence on all criteria. This confirms that the forming history influences the crash processing step considerably, and, thus, has to be taken into account. Furthermore, the sensitivity results confirm that, if the input distributions given by the forming history (cf. Figure 7.24) include only very small variations, these will have also minimal impact on the resulting distributions of the criteria considered in the crash process. This verifies that the tension-bearing test of the component behaves stable.

Almost all influential parameters have nonlinear behavior on average, which can be seen from the equal or higher values of $S2^j$ for each influential parameter in Figure 7.43.

| Class | Parameter | Damage % |
|---|---|---|
| TIClass$_1$ | $d_2$, $d_3$, $n$, $t$ | 65.19 |
| TIClass$_2$ | $d_1$ | 71.77 |
| TIClass$_3$ | $d_{\text{Shear1}}$, $R_0$, $R_{90}$, $K$ | 92.38 |
| TIClass$_4$ | $R_{45}$, $\varepsilon_0$ | 100 |

Table 7.10: Forming-to-crash: Parameter classification into total importance classes considering the crash processing step in timestep 90. The forming history is not taken into account. % denotes the average proportion of the overall variation explained by the union of the parameters up to the class considered.

Thus, only the total importance classes are listed exemplary for the criterion damage. The results of the classification based on the crash results considered as separate process are presented in Table 7.10. The results of the classification taking the history from the forming step into account are listed in Table 7.11. As already stated, the process parameters of the forming processing step influence the crash results substantially. Additionally, a comparison between the classification results for the criterion damage reveals that the parameter $d_1$ influencing the crash processing step considered separately, with almost seven percent, has much lower influence (1.5

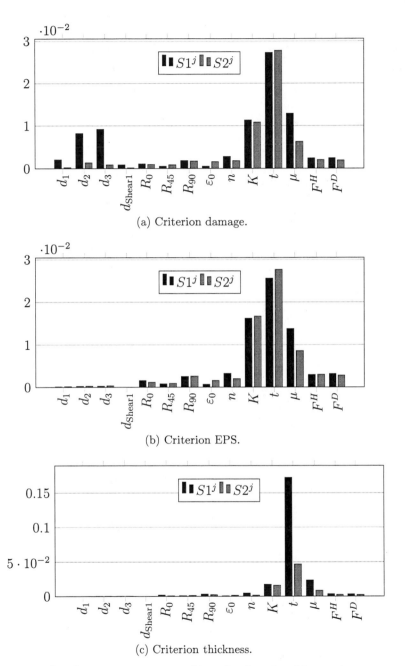

(a) Criterion damage.

(b) Criterion EPS.

(c) Criterion thickness.

Figure 7.43: Crash processing step taking the forming history into account: Linear $(S1^j)$ and nonlinear $(S2^j)$ accumulated sensitivity measures for the three criteria considered.

| Class | Parameter | Damage % |
|---|---|---|
| TIClass$_1$ | $K, \mu, t$ | 67.93 |
| TIClass$_2$ | $d_2, d_3, n, F^H, F^D$ | 91.21 |
| TIClass$_3$ | $R_{90}$ | 93.74 |
| TIClass$_4$ | $d_1, R_0, \varepsilon_0$ | 98.29 |
| TIClass$_5$ | $d_{\text{Shear1}}, R_{45}$ | 100 |

Table 7.11: Forming-to-crash: Parameter classification into total importance classes considering the crash processing step in timestep 90. The variations due to the forming history are included. % denotes the average proportion of the overall variation explained by the union of the parameters up to the class considered.

percent), when the forming history is taken into account. Hence, the assumption of 20 percent input variation of $d_1$ in the crash processing step, does not correspond to the realistically resulting variation in the criterion damage due to the forming history.

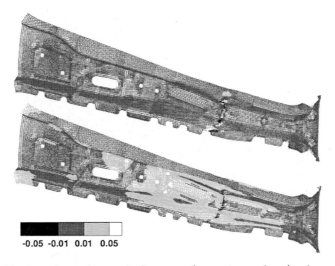

-0.05 -0.01 0.01 0.05

Figure 7.44: Local nonlinear influences (negative values) of parameter $d_3$ on the criterion damage. Comparison between the nonlinearity measure obtained for the crash processing step without (top) and with (bottom) taking the forming history into account, visualized on the final mesh of the nominal parameter set.

Furthermore, a comparison between the local distribution of the nonlinear influences of the important parameter $d_3$ on the criterion damage is presented in Figure 7.44. In both cases, the nonlinear influences, indicated with negative values, are high in the region in which a failure is initiated, if the B-pillar is overloaded unrealistically. Moreover, the figure shows that an additionally strong influence of the parameter $d_3$ arises in the region of a possible dent (light gray), when the forming history is taken into account.

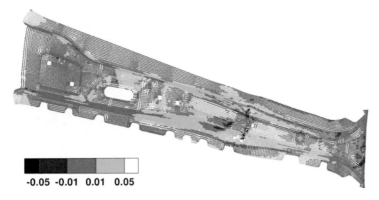

-0.05  -0.01  0.01  0.05

Figure 7.45: Local nonlinear influences of the forming process parameter friction on the criterion damage. The nonlinear influences, indicated with negative values, are particularly strong in the region in which a failure is initiated, if the B-pillar is overloaded unrealistically.

Large and partly nonlinear influences of the parameter friction are obvious in Figure 7.45 in the critical areas, if the B-pillar is overloaded unrealistically. The local distribution of the influences of the friction coefficient on the criteria EPS and thickness show similar results, and, thus, are not shown here.

In order to assess the crash process over time and the force up to the formation of a crack in the unrealistically overloaded B-pillar, force-displacement curves are usually analyzed. The force-displacement curves resulting from the ensemble of 2Npar + 1 simulation runs are shown in Figure 7.46. In detail, the curves obtained by the crash simulations considered separately are compared with the curves obtained by the crash simulations taking the forming history into account. Additionally, the force-displacement curves are clustered according to the assignment of the parameters to the total importance classes.

The curves show a high variation range, whether the forming history is taken into account or not. However, there is no sudden decrease in the force visible, that is, there is no crack initiated, when the forming history is not considered, except for $d_3 + 20\%$. Whereas, the initiation of a crack is clearly visible by the strong decrease in force, when the forming history is

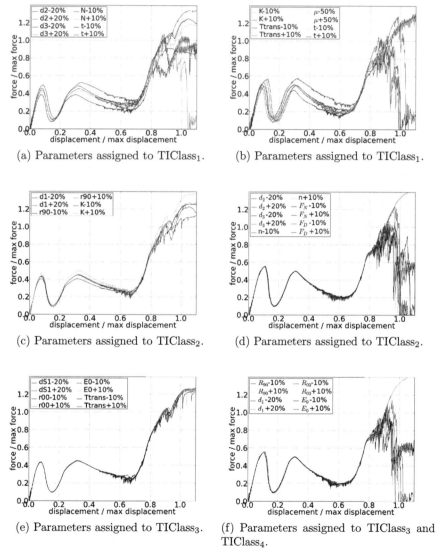

(a) Parameters assigned to TIClass$_1$.    (b) Parameters assigned to TIClass$_1$.

(c) Parameters assigned to TIClass$_2$.    (d) Parameters assigned to TIClass$_2$.

(e) Parameters assigned to TIClass$_3$.    (f) Parameters assigned to TIClass$_3$ and TIClass$_4$.

Figure 7.46: Comparison of force-displacement curves resulting from crash simulations with varied parameters. The history of the forming step is not taken into account (left), and is taken into account (right).

taken into account. Afterwards, a rather chaotic behavior of the curves is observed, which is not relevant anymore for the assessment of the result.

Moreover, the figure clearly shows that the first cluster provides a high variation, the second cluster already shows a strong tendency in the force-displacement curves, and the third cluster almost results in a single curve (at least up to the formation of a crack). Specifically, the force-displacement diagrams according to the parameter classification reflect the fact that the parameters assigned to $TIClass_1$ represent the strongest influences due to parameter variations.

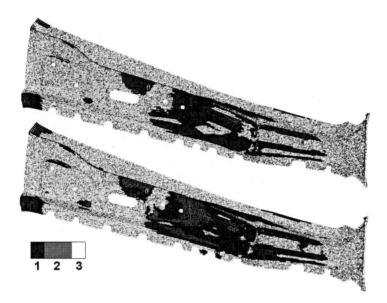

Figure 7.47: Comparison between the clustering results with Nclust $= 2$ (top) and Nclust $= 3$ (bottom) clusters based on the influences of variations of the parameter $d_3$ on the criterion damage. Local effects, like a dent (gray cluster) and a crack (black cluster) of the unrealistically overloaded B-pillar can be separated selecting three clusters.

We perform a clustering based on the nonlinear influences of parameter $d_3$ on the criterion damage taking the forming history into account, as derived in Section 4.2. We compare the results using Nclust $= 2$ clusters with the ones using Nclust $= 3$ clusters in Figure 7.47. When the number of clusters is set to three, the regions of important local effects could be separated. The cluster $CL_2$ corresponds to the local region in which a dent may occur, whereas the cluster $CL_1$ predominantly contributes to the region in which a crack might arise, if the B-pillar is overloaded unrealistically.

In summary, the strong influences of the process parameters of the forming process on the results of the crash process confirm that the consideration of the process history is essential in order to obtain realistic results in the last processing step.

## 7.3.5    Forecast Model Taking the Forming History Into Account

We analyze the entire process chain from forming to crash exemplary for the criterion damage. We aim at predicting the criterion damage resulting from the crash processing step by means of the extended and accelerated metamodel introduced in Section 6.1. Especially, possible failures arising if the B-Pillar is overloaded unrealistically should be predicted properly.

For this purpose, we consider exemplary the scenario in which only the four damage parameters $d_1, d_2, d_3$ and $d_{\text{Shear1}}$ are considered as random, all other parameters involved are assumed to be constant. This scenario has already been analyzed in Section 7.3.3 for the first processing step.

We take the history of the process, i.e., the forming process, into account. The analysis of the forming processing step has identified the two parameters $d_2$ and $d_3$ as the most influencing, contributing to 85 percent of the overall variation on average. The parameter classification of the crash processing step taking the forming history into account also assigns these two parameters to the first total importance class. Hence, these two parameters remain implicitly in the crash forecast model. That is, all resulting variations in the criterion damage from the forming history caused by variations of these two damage parameters are mapped to the crash processing step. A forecast model of the criterion damage in the crash step, including the variations of these two parameters, is expected to explain 85 percent of the overall variation on average (cf. Section 7.3.4). The local distribution of the proportion explained by the parameters $d_2$ and $d_3$ is shown in Figure 7.28.

In order to take the complete history of the process into account, all relevant scatter information, that is, the local distributions of thicknesses, EPS, and damage, is mapped to the crash processing step according to the mapping procedure derived in Section 6.3. To be more specific, first, the processed database containing distributions of the criteria resulting from the ensemble of forming simulation runs corresponding to the Nexp = 20 sampling points are mapped to the crash mesh. Then, an ensemble of crash simulation runs is performed using these resulting distributions of the criteria of the forming process explicitly as input distributions. Hence, the variations involved in the crash processing step are given by the resulting distributions of the criteria of the forming process. Finally, the accelerated RBF metamodel with Nexp = 20 samples is setup as forecast model **B-pillar-damage**, as in the first processing step. In this step, the procedure developed to deal with deleted mesh elements is applied.

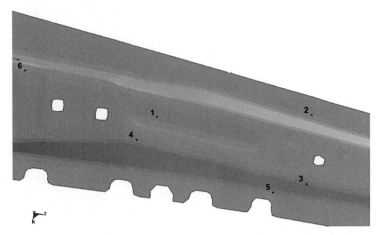

Figure 7.48: Six selected mesh nodes on the B-pillar in order to analyze the local process behavior (detail view).

We use the same five randomly selected out-of-sample points as in the prediction of the criterion damage in the forming processing step in order to validate the predicted results. That is, the distributions of the criteria resulting from the forming process are mapped to the crash process for each test point, and a corresponding crash simulation is performed. Recall that the full uncertainty caused by the four damage parameters considered in this model are represented by the test points. Analogous to the forming step, we specify six mesh nodes in order to analyze the local process behavior as well as the local prediction accuracy. These test nodes, highlighted in Figure 7.48, are specified in regions in which the criterion damage becomes possibly large.

As first scenario, a state after the formation of a crack is investigated. The distribution of the local absolute prediction errors **diff**$_{abs}$ according to Equation (6.2) in the test points selected, approximated with q-quantiles using the order statistics, is graphed in Figure 7.49. This shows that the absolute prediction error is below 0.002 on average. Moreover, the absolute prediction error is below 0.008 with 95 percent probability, which corresponds to ten percent relative to the reference solution.

In order to assess the results of the crash processing step, we investigate the prediction of possible failures, that is, the threshold $n_{\mathrm{del},\mathbf{M}}(i)$ derived in Equation (6.8), numerically. The resulting regions of failure predicted with different thresholds are compared in Figure 7.50. A failure will occur in the

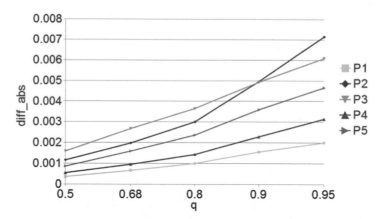

Figure 7.49: Approximated distribution function of absolute prediction errors in the five test points using B-pillar-damage-small.

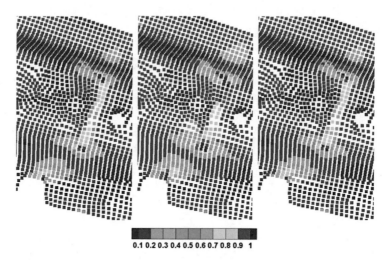

Figure 7.50: Comparison between the failure prediction with different thresholds (the damage values are scaled to $[0, 1]$ and shown on the initial, not deformed mesh). Failure occurs in the simulation (right), if the damage criterion exceeds the value 1. The prediction of the failure with the forecast model gives good results using the threshold derived (left). If the minimum of the contributing critical values is chosen as threshold, the failure criterion would be overestimated (center).

simulation, if the scaled criterion damage exceeds the value 1. This failure is predicted well by the forecast model using the threshold suggested, given by the median of the contributing thresholds to the setup of the metamodel. If the minimum of the contributing thresholds to the setup of the metamodel is specified as threshold, the failure criterion would be overestimated. In particular, the crack has progressed too far from both sides to the middle of the B-pillar.

Figure 7.51: Approximated distribution of the criterion damage (top, scaled to [0, 1]) together with local relative prediction errors (bottom) in test point $P_1$ in a state after the failure has been initiated (detail view).

Hence, we use the threshold suggested and review the local distribution of the criterion damage, specifically, the region in which a crack will occur with high probability, if the B-pillar is overloaded correspondingly. The predicted distribution of the criterion damage is exemplified in Figure 7.51 for test point $P_1$ in a state after a failure has been initiated. Nodes colored in red are identified with Equation (6.8) as nodes belonging to deleted elements, that is, in this regions a failure is predicted. Additionally, higher damage values arise in the local region where the punch force is applied directly (highlighted). Furthermore, the figure shows the local relative prediction error in this test point, which is very small almost everywhere. Few nodes with a higher prediction error are visible (red) only in the area of the predicted crack. In these nodes, the damage values are overestimated.

As second scenario, a forecast model over time is generated, in order to investigate the progress of the crash processing step. In this case, we use the database $\mathbf{M}$ defined in Equations (2.2), (2.3) in the setup phase of the metamodel. Since only the criterion damage is considered, we set Ncrit = 1. We evaluate the process up to the complete formation of a crack, that is, Nts = 90 states are taken into account. This leads to a huge amount of data, the number of rows of the matrix is Ndata > 7.5 millions. Nevertheless, since the number of columns Nexp = 20 is small, and the metamodel is sufficiently accelerated by a SVD, a complete approximation over all states and a prediction of the behavior of new designs becomes efficiently possible with the new PRO-CHAIN methodology. This allows also an approximation of the coordinates in each state. Hence, the deformation of the component under loading can be predicted.

For example, a complete prediction of a single test point, consisting of the deformations and the distribution of the criterion damage in each state, is predicted by the forecast model described in approximately five seconds on a standard 3 GHz Linux PC. The resulting deformation and distribution of the criterion damage for a test point is exemplified in Figure 7.52 for an early state and a state after a failure has been initiated. The geometry shows almost no difference to the simulation result, thus, the local differences are not shown here.

In conclusion, the deformed geometry is approximated with high accuracy in each state. The forecast model to predict the criterion damage is much better than the expected 85 percent of variation explained on average. To be more specific, the relative error is smaller than ten percent also with 95 percent probability. Moreover, the region and the size of the fracture and the crack are approximated properly. Hence, the crash processing step is represented well in its entire progress up to the final formation of a crack.

This high-quality metamodel derived is now evaluated in order to compute statistics of the criterion damage, which can be used in a subsequent robust optimization. The computation of $q$-quantiles is performed locally on the entire mesh and in each state as described in Remark 7.11 using Npred = 10000 samples.

**Remark 7.13.** *The approximation of the distribution function of the geometry, i.e., the deformations, is also possible. However, there is almost no difference to the simulation results, and, hence, this is not shown here.*

The local process behavior in single mesh nodes can be followed by the local approximation of the probability distribution function. Figure 7.53

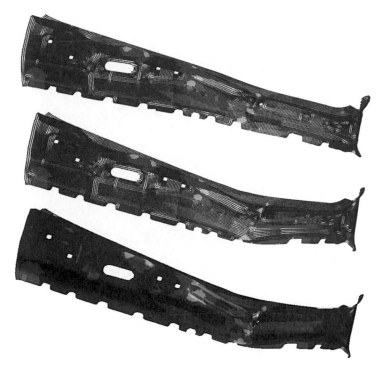

Figure 7.52: Prediction of the deformations by means of the metamodel in an early state A (top) and a state B after a failure has been initiated (center). For comparison, the simulation result in state B (bottom) is shown.

shows the approximated distribution of the criterion damage in the test nodes $N_4$ and $N_5$ in state ts $= 90$ after a crack has been initiated due to the unrealistic overload of the B-pillar. In particular, the figure shows that the damage values of $N_5$ are already above the computed threshold beginning with the 0.1-quantile. Thus, the test node $N_5$ belongs to failed elements with more than 90 percent probability. Whereas the damage values of $N_4$ do not reach the critical value at all, that is, in this local region no failure is expected.

The median, the 0.1-quantile, and the 0.9-quantile are presented in Figure 7.54 for state ts $= 50$ and state ts $= 90$, shown on the initial, not deformed mesh. In state ts $= 50$, the 0.1-quantile already shows the initiation of the fracture (red). A difference in the quantiles can be seen in the region where the punch force is applied directly. There, the 0.9-quantile shows higher damage values, which correspond to a dent in the component and may cause a failure in further timesteps. In state ts $= 90$, the fracture initiation from the side of the B-pillar and the crack (red) is clearly visible

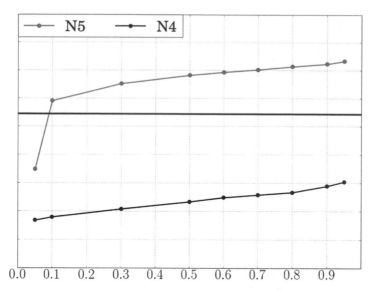

Figure 7.53: Approximated probability distribution function of the criterion damage in the test nodes $N_4$ and $N_5$. The black line corresponds to the threshold developed, which indicates that a node is predicted to belong at least to one deleted mesh element.

in all quantiles shown. The 0.9-quantile shows that the crack is progressed almost completely from one side to the other. Furthermore, higher damage values appear in the region of a possible dent (orange) in the 0.9-quantile. That is, with ten percent probability the damage values exceed these values shown, which may cause a failure in this region.

## 7.3.6   Conclusions

We have investigated the forming-to-crash process chain in detail. The parameter classification procedure has revealed the most influential parameters on the criteria considered. Additionally, it has detected local effects, like nonlinearities. Specifically, regions in which a dent and a crack might occur, if the B-pillar is overloaded absolutely unrealistically, have been identified successfully. This confirms that a local consideration and evaluation of the processes is essential in order to generate high-quality forecast models.

The comparison of the parameter classification results obtained by analyzing the second processing step separately, with the ones taking the history of the process into account, confirms the high relevance of the forming process parameters to the crash processing step. Moreover, it provides realistic variation ranges for the second processing step.

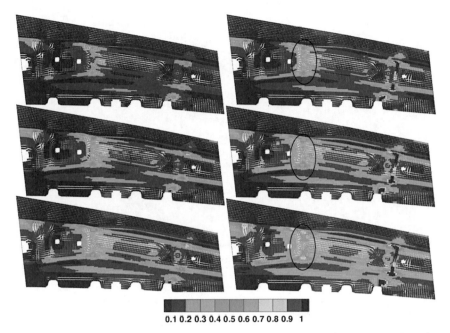

0.1 0.2 0.3 0.4 0.5 0.6 0.7 0.8 0.9  1

Figure 7.54: 0.1-quantile, median, and 0.9-quantile (top to bottom) of the damage distribution in state 50 (left) and state 90 (right) using B-pillar-damage-small (detail view). The median is scaled to $[0, 1]$, the other quantiles are scaled relative to the maximal value of the median.

An experimental validation of the results has been performed by a bending-tension test in [47]. A comparison between the experiment and simulations with and without taking the forming history into account is given in Figure 7.55. The figure shows that no failure will arise, if the influences from the deep drawing process are not taken into account. Especially, this demonstrates that only with the complete propagation of local thicknesses, EPS and damages from the forming process to the crash process the initiation and position of the crack can be simulated correctly.

Moreover, a comparison of the local distribution of the criterion damage from a crash variation with and without taking the forming history into account is presented in Figure 7.56. For comparison, the results are shown on the initial crash mesh. Taking the forming history into account an additional dent is strongly visible. This shows, all possible effects can only be reproduced correctly in a crash simulation, if the full history of the process, including all influences due to parameter uncertainties, is considered.

In conclusion, the propagation of all relevant scatter information from the previous forming step to the crash step improves the quality of the crash simulations, and, thus, the results of the overall process, considerably.

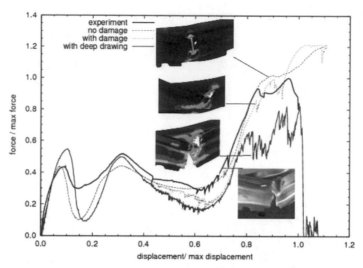

Figure 7.55: Comparison between the experimental result and simulations of the bending-tension test with and without taking the forming history into account (from [47]). The crack is caused by an absolutely unrealistic overload of the B-pillar.

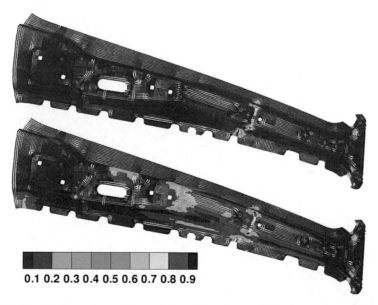

Figure 7.56: Comparison of resulting distribution of the criterion damage (scaled to $[0, 1]$) from a crash simulation varying the damage parameter $d_3$ plus 20 percent without (top) and with taking the forming history into account (bottom).

Furthermore, we have demonstrated that the accelerated RBF metamodels provide fast and high-quality forecast models for both processing steps. In particular, possible failures of the component under unrealistic overloading can be predicted properly. Moreover, the parameter classification procedure developed provides a good estimator of the average proportion of variation which is expected to be explained by the model after the parameter reduction is performed. Therefore, the forecast models, based on the iteratively extended and compressed database, can be set up automatically with a user-controlled accuracy.

Finally, we investigate the computational effort for the approximation of the probability distribution function of the criteria resulting from the final processing step of the forming-to-crash process chain considered. For this purpose, a state-of-the-art quasi-MC simulation is compared with the new PRO-CHAIN methodology developed.

For comparison, the memory requirement refers to all quantities which need to be stored in order to reconstruct the entire probability information. In both methods, the $P^2$-algorithm, introduced in Section 6.2, can be used to approximate the $q$-quantiles of the distribution of the criteria considered. However, in the case a quasi-MC simulation is used, all samples have to be stored in order to reconstruct these information. If the PRO-CHAIN methodology is used, the entire probability distribution function and new designs can be reconstructed from the processed databases stored.

| | Quasi-MC | PRO-CHAIN | Reduction |
|---|---|---|---|
| Memory | 2000 GB | 174(54) GB | 11.5(37) |

Table 7.12: Comparison of the memory requirements to approximate the probability distribution function of the criterion EPS between a quasi-MC method and the PRO-CHAIN methodology using `B-pillar-eps-big`. Brackets: acceleration by a SVD retaining $k = 26$ singular values.

The memory requirements for both methods are exemplified in Table 7.12 for the criterion EPS. The memory needed to store the forming simulation result, or the crash results with all timesteps, including all criteria is approximately 1 GB each. When the quasi-MC method is used, a large number of samples is required in each processing step in order to approximate the probability distribution function of the criteria. If a quasi-MC simulation with Npred = 1000 samples from the Halton sequence (cf. Section 3.1) is used in each step, 2000 GB memory will be required in total. Note that the number of samples depends on the number of parameters involved, thus, even much more samples may be needed.

When the new PRO-CHAIN methodology is used, the number of simulation runs required is minimized by an iterative extension of the database.

For example, we have observed in Section 7.3.3 that Nexp $= 87$ samples are sufficient to construct an appropriate forecast model for the criteria EPS and thickness. Thus, in each processing step Nexp $= 87$ samples are used, which leads to a memory requirement of 174 GB. Additionally, we have shown that a high compression factor can be achieved by applying a SVD to the database. In the case of the model B-pillar-eps-big, it has been sufficient to retain $k = 26$ singular values. If this compression is applied in each processing step, the overall required memory will reduce to 54 GB. In total, a substantial memory reduction has been achieved in this example.

| Computational time | Quasi-MC | PRO-CHAIN | Speed-up |
|---|---|---|---|
| 1 simulation at a time | 1.48 years | 47.1 days | 11.5 |
| 40 simulations in parallel | 13.5 days | 1.18 days | 11.5 |

Table 7.13: Comparison of the computational time to approximate the probability distribution function of the criterion EPS between a quasi-MC method and the PRO-CHAIN methodology using B-pillar-eps-big.

Furthermore, we compare the computational time of both methods to approximate the probability distribution function of the final criterion EPS, given in Table 7.13. The $P^2$-algorithm is used in both cases to approximate nine quantiles with $q \in [0.05, 0.95]$. In the forming processing step, only the final distribution of the criterion, i.e., the last timestep, is approximated.

A single forming simulation run takes approximately 10 h performed on a Linux 2.6 GHz HPC Cluster using 16 compute nodes, and a crash simulation run takes approximately 3 h, respectively. The prediction of each criterion considered, by means of the accelerated RBF metamodel developed, takes approximately 0.007 s per criterion and timestep in each processing step.

As already mentioned, we perform a quasi-MC simulation with Npred $= 1000$ samples in each processing step. The time for the $P^2$ algorithm is several minutes, and, thus, negligible compared with the simulation runtime.

In the case of the PRO-CHAIN methodology, we use the forecast model B-pillar-eps-big. Thus, Nexp $= 87$ simulation runs have to be performed in each processing step, which dominates the computational time. Then, the metamodels are evaluated with Npred $= 10000$ samples in each processing step in order to approximate the quantiles aimed at. The time for the construction of the metamodel and the SVD of the database, together with the evaluation of the metamodel is also several minutes, and, thus, negligible compared with the simulation runtime. In total, an enormous speed-up of 11.5 is achieved for this example using the new methodology developed compared with a state-of-the-art quasi-MC method.

In conclusion, a high performance gain and memory reduction can be achieved by the PRO-CHAIN methodology. This is reached by the dimension reduction and iterative processing of the database derived, while the accuracy of the results is kept comparable with the simulation results. This high speed-up and the low memory requirements allow the efficient propagation of all relevant variations from one processing step to the next. Finally, this enables the efficient approximation of the probability distribution function, which is essential to achieve a robust optimization of the component considered.

# Chapter 8

---

# Conclusions and Future Directions

---

Product components are usually run through several processing and analysis steps in the product development process. Each processing step commonly involves many parameters, which are subject to variations. These may have a substantial, and also often very local, impact on the result. Thus, a local analysis and quantification of uncertainties is important, particularly when considering whole process chains. The history of the process, including the impact of these uncertainties, has to be taken into account in subsequent processing steps in order to improve the final result substantially. However, the consideration of the history of the process including all relevant variations, is still not state-of-the-art.

Moreover, due to an ensemble of simulation runs with many simulation timesteps and hundred thousands of nodes for a single component, many challenges due to this huge amount of high dimensional data have to be met. The locality of important physical effects makes a local analysis unavoidable. Together with the curse of dimensionality present in high dimensional data, this has to date made an efficient analysis of uncertainties for whole process chains impossible.

In this work, we have developed the **PRO-CHAIN methodology** for the efficient analysis of process chains involving parameter uncertainties. In particular, we have demonstrated that the consideration of variations, arising from the process history, within the final processing step improves the forecasting quality considerably.

State-of-the-art parameter sensitivity analysis approaches are usually based on global methods and fail to identify important local influences. Additionally, variance-based sensitivity methods are very time-consuming.

Hence, these methods are not directly applicable in industrially relevant problems, but have to be based on appropriate metamodels.

To address this, we have developed a **parameter classification approach** using a local sensitivity analysis that **fully automatically** classifies the influences of parameter variations on the criteria considered. Moreover, the approach proposed computes sensitivity measures locally, which provides a **local estimate of the average prediction quality** of a forecast model taking the uncertainties of corresponding parameters into account.

The procedure developed minimizes the number of simulation runs required, so that it may employ simulation results directly. Furthermore, based on the results of the parameter classification, the parameter space is reduced in order to minimize the effort of the remaining analysis steps. Subsequent metamodels can be constructed with the resulting, reduced number of random parameters. The clustering approach proposed, based on the results of the local nonlinearity measure, can further improve the forecast quality of the metamodels locally by applying more advanced approximation methods in the regions of interest. We have demonstrated the efficiency and benefit of this parameter classification procedure for several industrial applications.

The PRO-CHAIN methodology introduces an iterative procedure for the extension of the database in the case of nonlinear parameters to ensure a suitable accuracy. Additionally, the ensemble of simulation runs is compressed by a singular value decomposition (SVD) leading to a further reduction of the computational complexity. This processing of the database is an essential step within the methodology proposed to partly avoid the problems due to the high dimensionality.

We have accelerated radial basis function (RBF) metamodels using the SVD of the database. This allows a **fast and accurate prediction** of new designs fully locally. Specifically, local effects due to a certain change in the parameter set can be quantified directly. Moreover, we have generalized these forecast models in order to enable the prediction of failure initiation in crash processes. That is, a procedure to deal with deleted mesh elements and to predict failure locally has been introduced. This is an essential feature to assess the results of crash processes. In conclusion, the metamodels proposed allow an immediate answer to what-if scenarios avoiding additional time-consuming simulation runs. We have investigated an industrial bending tension test of a B-pillar for validation of the developed methods. This shows that the new forecast models allow a fast and accurate prediction of the behavior of new designs including the local distributions of relevant design criteria. In particular, important influences have been identified and the local behavior has been characterized. Moreover, the failure initiation has been predicted successfully.

A meaningful statistical analysis of whole process chains requires the variations of all relevant parameters to be taken into account locally. The standard approach for statistical analysis is still a (quasi-) Monte Carlo method. Due to the very high number of samples required to get suitable statistical information, this approach is not appropriate to be used directly with complex simulation runs. We have derived an efficient way to **approximate quantiles** of the probability distribution function of the criteria considered by a local Monte Carlo evaluation of the high-quality metamodels constructed. This allows the inclusion of, for example, the median and additional quantiles as robustness measures in a subsequent optimization task. If nonlinear parameters are involved, this is expected to give much better results than the consideration of standard robustness measures, such as the mean and standard deviation. Additionally, the statistical information is provided locally on the entire mesh, which is more meaningful than global measures, particularly for complex nonlinear processes. We have illustrated the high decrease of computational time with the same precision compared with state-of-the-art Monte Carlo simulations by means of an industrial forming application.

In combination, the developed parameter classification procedure and the accurate forecast models enable the **propagation of essential scatter** due to parameter variations locally from one processing step to the next. That is, with the methodology newly developed it becomes possible to take not only the history of the process into account in the last processing step, but also the impact of all relevant variations can be considered locally on the entire mesh.

Moreover, we have derived an **estimator of the average approximation error** in a predicted new design in a single processing step, which is computed directly within the new methodology. An additional **estimator** of the **maximal approximation error** for the entire process chain has been derived theoretically.

A complex forming-to-crash process chain investigated has demonstrated that a **considerably better forecasting quality** of the crash process is achieved using the PRO-CHAIN methodology. Furthermore, a **memory reduction** of 37, and a **speed-up** of 11.5 achieved in this industrial application has demonstrated that the enormous performance gain by the new methodology allows the efficient propagation of all relevant variations within a whole process chain.

Bringing all individual components of the newly developed PRO-CHAIN methodology together, we have created a basis for a virtual robust design process for whole process chains. The benefit and efficiency of the methodology developed has been demonstrated using industrial applications from the automotive industry. Furthermore, a patent on this new methodology invented has been granted [114]. In summary, the methodology proposed, enables the **analysis, quantification and propagation of uncertain-**

**ties locally** on the entire mesh for complex process chains, which has until now not been feasible for reasons of cost and efficiency.

## Future Directions

We have demonstrated that the newly developed PRO-CHAIN methodology enables an efficient analysis and transfer of uncertainties over several steps of process chains, which leads to considerably improved prediction results of complex processing steps.

In future work, additional processes should be included in the forming-to-crash process chain. For example, parameter variations arising through the connection of several parts, as occurring in spot-welded joints, can influence the resulting load bearing capacity of the component strongly. Thus, the variations caused by joining processes may contribute strongly to the robustness of the final solution. Therefore, these processes should be part of the analysis.

Additionally, future directions will deal with the integration of the results obtained by the methodology newly developed, specifically, the approximated quantiles, into robustness measures and appropriate optimization criteria in order to provide a fully comprehensive analysis and optimization tool.

In the area of a broader range of applications, future studies will address the investigation of other industrially relevant applications and the consideration of complete vehicles with the PRO-CHAIN methodology proposed. Applications arise from, for example, the semiconductor industry, the mechanical engineering sector, and energy networks.

When considering complete vehicles, a huge amount of data has to be investigated. For example, using all time steps, which are usually several thousands in crash processes, and a set of car components, or a whole vehicle, millions of nodes have to be analyzed. Together, this results in matrices of the size millions times hundreds. Due to this fact, handling "big data" will become an important issue for the analysis of process chains.

"Analyzing large data sets – so-called big data – will become a key basis of competition, [...] growth [and] innovation" for companies, as a report of McKinsey Global Institute investigated in 2011 [78].

Big data cannot be analyzed efficiently with state-of-the-art data analysis tools. Due to this reason, new big data analysis tools are being developed. One of these frameworks to analyze big data on cluster computers is *Apache Hadoop*, a java-based open-source framework "for the distributed processing of large data sets across clusters of computers" [2]. Details on Hadoop can be found in [3].

As we have demonstrated, the prediction of the behavior of new designs by means of a metamodel accelerated by a SVD of the database used is an essential step within the PRO-CHAIN methodology. To accelerate the prediction of out-of-sample points, and, therewith also the computation of statistics, the SVD can be carried out within a Hadoop framework. This parallel and scalable computation of the SVD makes this PRO-CHAIN step possible also with big data.

An application of Hadoop for the analysis of big data with a SVD is evaluated in the bachelor thesis [68]. An important result of the thesis has pointed out that the Hadoop implementation of a SVD is beneficial for matrices exceeding the size of one million times 200 compared with a standard C++ implementation.

In summary, these first results show that a combination of a Hadoop big data solution and parts of the PRO-CHAIN methodology is possible. We have shown in this work that other parts of the new methodology can also be parallelized. Thus, we are confident that additional PRO-CHAIN parts are suitable for Hadoop, which has to be analyzed in future work.

# Bibliography

[1]  S. Ackermann, L. Gaul, M. Hanss, and T. Hambrecht. Principal component analysis for detection of globally important input parameters in nonlinear finite element analysis. In *Weimar Optimization and Stochastic Days 5.0*, 2008.

[2]  Apache Software Foundation. `http://hadoop.apache.org/`, 2013. Retrieved 31 July 2013.

[3]  Apache Software Foundation. Hadoop 1.2.1 Documentation. `http://hadoop.apache.org/docs/stable/`, 2013. Retrieved 31 July 2013.

[4]  I. M. Babuška, F. Nobile, and R. Tempone. A stochastic collocation method for elliptic partial differential equations with random input data. *SIAM Review*, 52(2):317–355, 2010.

[5]  D. Banabic. *Sheet Metal Forming Processes - Constitutive Modelling and Numerical Simulation*. Springer, 2010.

[6]  B.-A. Behrens and E. Doege. Erhöhung der Umformgrenzen beim Tief- und Streckziehen durch lokal vorstrukturierte Platinen. Technical report, DFG Forschungsbericht Do 190/138-4, 2004.

[7]  M. Belkin and P. Niyogi. Laplacian eigenmaps for dimensionality reduction and data representation. *Neural Computation*, 15:1373 – 1396, 2003.

[8]  R. Bellman. *Adaptive Control Processes: A Guided Tour*. Princeton University Press, 1961.

[9]  M. W. Berry. Large scale sparse singular value computations. *International Journal of Supercomputer Applications*, 6(1):13–49, 1992.

[10]  H.-G. Beyer and B. Sendhoff. Robust optimization - a comprehensive survey. *Computer Methods in Applied Mechanics and Engineering*, 196:3190–3218, 2007.

[11] M. Bieri and C. Schwab. Sparse high order FEM for elliptic sPDEs. *Computer Methods in Applied Mechanics and Engineering*, 198(13-14):1149–1170, 2009.

[12] D. Borsotto, T. Clees, I. Nikitin, L. Nikitina, D. Steffes-lai, and C.-A. Thole. Sensitivity and robustness aspects in focused ultrasonic therapy. In *Proceedings of the 3rd International Conference on Engineering Optimization (EngOpt), Rio de Janeiro, Brazil*, 2012.

[13] M. D. Buhmann. *Radial Basis Functions: Theory and Implementations*. Cambridge University Press, 2003.

[14] G. van Bühren, N. Hornung, T. Clees, and L. Nikitina. Aspects of adaptive hierarchical RBF meta-models for optimization. *Journal of Computational Methods in Sciences and Engineering*, 12:5 – 23, 2012.

[15] H.-J. Bungartz and M. Griebel. Sparse grids. *Acta Numerica*, 13:1 – 123, 2004.

[16] R. E. Caflisch. Monte Carlo and quasi-Monte Carlo methods. *Acta Numerica*, 7:1 – 49, 1998.

[17] F. Camastra and A. Vinciarelli. Intrinsic dimension estimation of data: an approach based on Grassberger - Procaccia's algorithm. *Neural Processing Letters*, 14:27 – 34, 2001.

[18] L. Cizelj, B. Mavko, and H. Riesch-Oppermann. Application of first and second order reliability methods in the safety assessment of cracked steam generator tubing. *Nuclear Engineering and Design*, 147:359–368, 1994.

[19] T. Clees, I. Nikitin, L. Nikitina, and S. Pott. Efficient quantile estimators for river bed morphodynamics. In *Proceedings of the 3rd International Conference on Simulation and Modeling Methodologies, Technologies and Applications (SIMULTECH)*. SCITEPRESS, 2013. ISBN:978-989-8565-69-3.

[20] T. Clees, I. Nikitin, L. Nikitina, and C.-A. Thole. Nonlinear metamodeling and robust optimization in automotive design. In *Proceedings of the 1st International Conference on Simulation and Modeling Methodologies, Technologies and Applications (SIMULTECH)*, 2011.

[21] T. Clees, I. Nikitin, L. Nikitina, and C.-A. Thole. Analysis of bulky crash simulation results: Deterministic and stochastic aspects. In N. Pina, J. Kacprzyk, and J. Filipe, editors, *Simulation and Modeling Methodologies, Technologies and Applications*, volume 197 of *Advances in Intelligent Systems and Computing*, pages 225–237. Springer Berlin Heidelberg, 2013.

[22] T. Clees and D. Steffes-lai. PRO-CHAIN: Efficient statistical analysis of process chains. *ERCIM News*, 81:29 – 30, 2010.

[23]  T. Clees and D. Steffes-lai. PRO-CHAIN: Efficient statistical analysis of process chains applied to a forming-to-crash example. In *Proceedings of the 9th LS-DYNA Users' Conference, Bamberg, Germany*, 2010.

[24]  T. Clees, D. Steffes-lai, M. Helbig, K. Roll, and M. Feucht. Process chain forming to crash: Efficient stochastic analysis. In *Proceedings of the 7th European LS-DYNA Users' Conference, Salzburg, Austria*, 2009.

[25]  T. Clees, D. Steffes-lai, M. Helbig, and D.-Z. Sun. Statistical analysis and robust optimization of forming processes and forming-to-crash process chains. *International Journal of Material Forming*, 3 (Supplement 1):45 – 48, 2010.

[26]  V. Costanza and J. H. Seinfeld. Stochastic sensitivity analysis in chemical kinetics. *Journal of Chemical Physics*, 74:3852 – 3858, 1981.

[27]  H. A. David and H. N. Nagaraja. *Order statistics*. John Wiley & Sons, 3rd edition, 2003.

[28]  P. Demartines and J. Herault. Curvilinear component analysis: A self-organizing neural network for nonlinear mapping of data sets. *IEEE Transactions on Neural Networks*, 8(1):148 – 154, 1997.

[29]  E. W. Dijkstra. A note on two problems in connexion with graphs. *Numerische Mathematik*, 1:269 – 271, 1959.

[30]  D. L. Donoho. High-dimensional data analysis: The curses and blessings of dimensionality. In *Proceedings of the AMS Conference "Math Challenges of the 21st Century"*, 2000.

[31]  C. Eckart and G. Young. The approximation of one matrix by another of lower rank. *Psychometrika*, 1(3):211 – 218, 1936.

[32]  H. Fang, M. Rais-Rohani, Z. Liu, and M. F. Horstemeyer. A comparative study of metamodeling methods for multiobjective crashworthiness optimization. *Computers and Structures*, 83:2121–2136, 2005.

[33]  G. S. Fishman. *Monte Carlo: concepts, algorithms and applications*. Springer, New York, USA, 1996.

[34]  Fraunhofer SCAI. *DesParO User's Manual and Release Notes*, 2.1 edition, 2011.

[35]  Fraunhofer SCAI. *SCAIMapper Documentation*, 1.1.2 edition, January 2011.

[36]  K. Fukunaga and D. R. Olsen. An algorithm for finding intrinsic dimensionality of data. *IEEE Transactions on Computers*, C-20(2):176 – 193, 1971.

[37]  T. Gerstner and M. Griebel. Numerical integration using sparse grids. *Numerical Algorithms*, 18:209–232, 1998.

[38]  R. G. Ghanem and P. D. Spanos. *Stochastic finite elements: a spectral approach*. Springer, New York, USA, 1991.

[39]  V. Gödel, M. Merklein, and B. Oberpriller. Influence of the variation of material properties on the risk of failure in dependency of the materials flow condition during forming processes. *International Journal of Material Forming*, 3 (Supplement 1):101 – 104, 2010.

[40]  G. H. Golub and C. Reinsch. Singular value decomposition and least squares solutions. *Numerische Mathematik*, 14:403 – 420, 1970.

[41]  G. H. Golub and C. F. Van Loan. *Matrix Computations*. J. Hopkins University Press, 3rd edition, 1996.

[42]  P. Grassberger and I. Procaccia. Measuring the strangeness of strange attractors. *Physica*, D9:189 – 208, 1983.

[43]  A. Griewank and A. Walther. *Evaluating Derivative, Principles and Techniques of Algorithmic Differentiation*, volume 2. SIAM, 2008. ISBN: 978-0-898716-59-7.

[44]  H. H. Harman. *Modern Factor Analysis*. The University of Chicago Press, Chicago, 3rd edition, 1976.

[45]  F. E. Harrell and C. E. Davis. A new distribution-free quantile estimator. *Biometrika*, 69:635 – 640, 1982.

[46]  R. J. Harris. *A primer of multivariate statistics*. Lawrence Erlbaum Associates, Inc., 3rd edition, 2001.

[47]  M. Helbig, F. Andrieux, D. Steffes-lai, and T. Clees. Development of a material model for the process chain forming to crash taking stochastic and deterministic influences into account. In *Proceedings of the ANSYS Conference & 27th CADFEM Users' Meeting, Leipzig, Germany*, 2009.

[48]  J. C. Helton and R. L. Iman. An investigation of uncertainty and sensitivity analysis techniques for computer models. *Risk Analysis*, 8(1):71 – 90, 1988.

[49]  J. C. Helton, J. D. Johnson, C. J. Sallaberry, and C. B. Storlie. Survey of sampling-based methods for uncertainty and sensitivity analysis. *Reliability Engineering and System Safety*, 91:1175 – 1209, 2006.

[50]  M. Hermle, M. Feucht, and T. Frank. Crashsimulation in der Fahrzeugentwicklung - Anforderungen an CAE und Software. In *LS-DYNA Anwenderforum, Filderstadt, Germany*, 2013.

[51]   T. Homma and A. Saltelli. Importance measures in global sensitivity anal-
       ysis of nonlinear models. *Reliability Engineering and System Safety*, 52:1 –
       17, 1996.

[52]   P. Hora, J. Heingärtner, N. Manopulo, L. Tong, D. Hortig, A. Neumann,
       and K. Roll. On the way from an ideal virtual process to the modelling of
       the real stochastic. In *Proceedings of the Forming Technology Forum, ETH
       Zurich (Switzerland)*, 2011.

[53]   J. E. Jackson. *A User's Guide To Principal Components*. J. Wiley & Sons,
       2003.

[54]   R. K. Jaiman, X. Jiao, P. H. Geubelle, and E. Loth. Conservative load
       transfer along curved fluid-solid interface with non-matching meshes. *Jour-
       nal of Computational Physics*, 218:372 – 397, 2006.

[55]   R. Jain and I. Chlamtac. The P2 algorithm for dynamic calculation of
       quantiles and histograms without storing observations. *Communications of
       the ACM*, 28(10):1076 – 1085, 1985.

[56]   A. Janon, M. Nodet, and C. Prieur. Confidence intervals for sensitivity
       indices using reduced-basis metamodels. *ArXiv e-prints*, 1102.4668, 2011.

[57]   X. Jiao and M. Heath. Common-refinement-based data transfer between
       non-matching meshes in multiphysics simulations. *International Journal
       for Numerical Methods in Engineering*, 61:2402 – 2427, 2004.

[58]   I. T. Jolliffe. *Principal Component Analysis*. Springer, 2nd edition, 2002.

[59]   I. T. Jolliffe and B. J. T. Morgan. Principal component analysis and ex-
       ploratory factor analysis. *Statistical Methods in Medical Research*, 1:69 –
       95, 1992.

[60]   F. Jurecka. *Robust Design Optimization based on metamodeling techniques*.
       PhD thesis, TU Muenchen, 2007.

[61]   N. Kambhatla and T. K. Leen. Dimension reduction by local principal
       component analysis. *Neural Computation*, 9:1493 – 1516, 1997.

[62]   T. Kanungo, D. M. Mount, N. S. Netanyahu, C. D. Piatko, R. Silverman,
       and A. Y. Wu. An efficient k-means clustering algorithm: analysis and
       implementation. *IEEE Transactions on Pattern Analysis and Machine In-
       telligence*, 24(7):881 – 892, 2002.

[63]   A. J. Keane and P. B. Nair. *Computational Approaches for Aerospace
       Design: The Pursuit of Excellence*. J. Wiley and Sons, 2005.

[64]   B. Kegl. Intrinsic dimension estimation using packing numbers. *Advances
       in Neural Information Processing Systems*, 15:681 – 688, 2002.

[65] J. P. C. Kleijnen. An overview of the design and analysis of simulation experiments for sensitivity analysis. *European Journal of Operational Research*, 164:287 – 300, 2005.

[66] J. R. Koehler and A. B. Owen. Computer experiments. In S. Ghosh and C. R. Rao, editors, *Handbook of statistics*, volume 13, pages 261 – 308. Elsevier, Amsterdam, 1996.

[67] T. Kohonen. *Self-Organizing Maps*. Springer, 1995.

[68] C. Kolberg. Nutzung von Apache Hadoop zur Analyse großer Datenbestände in einem Cluster-Computer. Bachelor thesis, Europäische Fachhochschule, 2013.

[69] J. B. Kruskal. Multidimensional scaling by optimizing goodness of fit to a nonmetric hypothesis. *Psychometrika*, 29(1):1 – 27, 1964.

[70] S. Kucherenko, M. Rodriguez-Fernandez, C. Pantelides, and N. Shah. Monte Carlo evaluation of derivative-based global sensitivity measures. *Reliability Engineering and System Safety*, 94:1135 – 1148, 2009.

[71] M. Ledoux. *The concentration of measure phenomenon*. AMS, 2009.

[72] J. A. Lee, A. Lendasse, and M. Verleysen. Nonlinear projection with curvilinear distances: Isomap versus curvilinear distance analysis. *Neurocomputing*, 57:49 – 76, 2004.

[73] J. A. Lee and M. Verleysen. *Nonlinear Dimensionality Reduction*. Springer, 2007.

[74] Livermore Software Technology Corporation, Livermore, California. *LS-DYNA Theory Manual*, March 2006.

[75] Livermore Software Technology Corporation (LSTC), Livermore, California. *LS-DYNA Keyword User's Manual*, 971 edition, 2007.

[76] G. J. A. Loeven, J. A. S. Witteveen, and H. Bijl. Probabilistic collocation: an efficient non-intrusive approach for arbitrary distributed parametric uncertainties. In *Proceedings of the 45th AIAA Aerospace Sciences Meeting, Reno, Nevada*, volume 6, pages 3845–3858, 2007.

[77] D. J. C. MacKay. *Information Theory, Inference, and Learning Algorithms*. Cambridge University Press, 2003.

[78] J. Manyika, M. Chui, B. Brown, J. Bughin, R. Dobbs, C. Roxburgh, and A. Hung Byers. Big data: The next frontier for innovation, competition, and productivity. http://www.mckinsey.com/insights/business_technology/big_data_the_next_frontier_for_innovation, May 2011. Retrieved 31 July 2013.

[79]  Z. Marciniak, J. L. Duncan, and S. J. Hu. *Mechanics of Sheet metal forming*. Butterworth-Heinemann, 2002.

[80]  D. C. Montgomery. *Design and Analysis of Experiments*. John Wiley & Sons, 4th edition, 1997.

[81]  W. J. Morokoff and R. E. Caflisch. Quasi-Monte Carlo integration. *Journal of Computational Physics*, 122:218 – 230, 1995.

[82]  T. Most and J. Will. Efficient sensitivity analysis for virtual prototyping. In J. Eberhardsteiner, H. J. Böhm, and F.G. Rammerstorfer, editors, *Proceedings of the European Congress on Computational Methods in Applied Sciences and Engineering (ECCOMAS), Vienna, Austria*, 2012. ISBN: 9783950353709.

[83]  H. Müllerschön, D. Lorenz, W. Roux, M. Liebscher, S. Pannier, and K. Roll. Probabilistic analysis of uncertainties in the manufacturing process of metal forming. In *Proceedings of the 6th European LS-DYNA Users' Conference, Sweden*, 2007.

[84]  R. H. Myers, D. C. Montgomery, and C. M. Anderson-Cook. *Response Surface Methodology: Process and Product Optimization Using Designed Experiments*. Wiley and Sons, 3rd edition, 2009.

[85]  H. Niederreiter. Quasi-Monte Carlo methods and pseudo-random numbers. *Bulletin of the American Mathematical Society*, 84:957 – 1041, 1978.

[86]  H. Niederreiter. *Random Number Generation and Quasi-Monte Carlo Methods*. SIAM, 1992.

[87]  I. Nikitin, L. Nikitina, and T. Clees. Stochastic analysis and nonlinear metamodeling of crash test simulation and their application in automotive design. In F. Columbus, editor, *Computational Engineering: Design, Development and Applications*. Nova Science, New York, 2011.

[88]  I. Nikitin, L. Nikitina, and T. Clees. Nonlinear metamodeling of bulky data and applications in automotive design. In M. Guenther, editor, *Progress in Industrial Mathematics at ECMI 2010*, volume 17 of *Mathematics in Industry*, pages 295 – 301. Springer, 2012.

[89]  F. Nobile, R. Tempone, and C. G. Webster. An anisotropic sparse grid stochastic collocation method for elliptic partial differential equations with random input data. *SIAM Journal on Numerical Analysis*, 46(5):2411–2442, 2008.

[90]  A. Oeckerath and K. Wolf. Improved product design using mapping in manufacturing process chains. In *Proceedings of the 9th LS-DYNA Users' Conference, Bamberg, Germany*, 2010.

[91]   W. H. Press, S. A. Teukolsky, W. T. Vetterling, and B. P. Flannery. *Numerical Recipes in C: The Art of Scientific Computing*. Cambridge University Press, 2nd edition, 1992.

[92]   H. Rabitz, M. Kramer, and D. Dacol. Sensitivity analysis in chemical kinetics. *Annual Review of Physical Chemistry*, 34:419 – 461, 1983.

[93]   R. Rackwitz. Reliability analysis – a review and some perspectives. *Structural Safety*, 23:365 – 395, 2001.

[94]   U. Reuter, Z. Mehmood, C. Gebhardt, M. Liebscher, H. Müllerschön, and I. Lepenies. Using LS-OPT for meta-model based global sensitivity analysis. In *Proceedings of the 8th European LS-DYNA Users' Conference, Strassbourg, France*, 2011.

[95]   B. Rhein, T. Clees, and M. Ruschitzka. Uncertainty quantification using nonparametric quantile estimation and metamodeling. In J. Eberhard-steiner, H. J. Böhm, and F. G. Rammerstorfer, editors, *Proceedings of the European Congress on Computational Methods in Applied Sciences and Engineering (ECCOMAS), Vienna, Austria*, pages 3694 – 3710, 2012. ISBN: 9783950353709.

[96]   J. A. Rice. *Mathematical statistics and data analysis*. Cengage Learning, 3rd edition, 2007.

[97]   K. Roll. Simulation of sheet metal forming - necessary developments in the future. In *Proceedings of the 7th International NUMISHEET Conference, Interlaken, Switzerland*, 2008.

[98]   J. Sacks, W. J. Welch, T. J. Mitchell, and H. P. Wynn. Design and analysis of computer experiments. *Statistical Science*, 4(4):409–423, 1989.

[99]   A. Saltelli, M. Ratto, T. Andres, F. Campolongo, J. Cariboni, D. Gatelli, M. Saisana, and S. Tarantola. *Global Sensitivity Analysis - The Primer*. John Wiley & Sons, 2008.

[100]  J. W. Sammon. A nonlinear mapping for data structure analysis. *IEEE Transactions on Computers*, C-18(5):401 – 409, 1969.

[101]  L. K. Saul and S. T. Roweis. Think globally, fit locally: unsupervised learning of low dimensional manifolds. *Journal of Machine Learning Research*, 4:119 – 155, 2003.

[102]  B. Schölkopf, A. Smola, and K.-R. Müller. Nonlinear component analysis as a kernel eigenvalue problem. *Neural Computation*, 10:1299 – 1319, 1998.

[103]  M. E. Sfakianakis and D. G. Verginis. A new family of nonparametric quantile estimators. *Communication in Statistics - Simulation and Computation*, 37:337 – 345, 2008.

[104] T. W. Simpson, J. D. Peplinski, P. N. Koch, and J. K. Allen. On the use of statistics in design and the implications for deterministic computer experiments. In *Proceedings of the ASME Design Engineering Technical Conferences, Sacramento, California*, 1997.

[105] D. Skillicorn. *Understanding Complex Datasets: Data Mining with Matrix Decompositions*. Chapman & Hall / CRC, 2007.

[106] S. A. Smolyak. Quadrature and interpolation formulas for tensor products of certain classes of functions. *Dokl. Akad. Nauk SSSR (Proceedings of the Russian Academy of Sciences)*, 148:1042 – 1045, 1963. [in Russian], Soviet Mathematics Doklady 4,240 - 243 [English transl.].

[107] I. M. Sobol. Global sensitivity indices for nonlinear mathematical models and their Monte Carlo estimates. *Mathematics and Computers in Simulation*, 55:271 – 280, 2001.

[108] I. M. Sobol and S. Kucherenko. Derivative based global sensitivity measures and their link with global sensitivity indices. *Mathematics and Computers .in Simulation*, 79:3009 – 3017, 2009.

[109] P. Spethmann, C. Herstatt, and S. H. Thomke. Crash simulation evolution and its impact on R&D in the automotive applications. *International Journal of Product Development*, 8(3):291 – 305, 2009.

[110] D. Steffes-lai. Echtzeit-Toleranzanalyse eines RBF-Metamodells mit Anwendung in der Fahrzeugentwicklung. Diplomarbeit, Mathematisches Institut der Universität zu Köln, 2007.

[111] D. Steffes-lai and T. Clees. Efficient stochastic analysis of process chains. *The International Journal of Multiphysics*, Special edition: Multiphysics simulations - Advanced methods for industrial engineering:231 – 238, 2010.

[112] D. Steffes-lai and T. Clees. Statistical analysis of process chains: Novel PRO-CHAIN components. In *Proceedings of the 8th European LS-DYNA Users' Conference, Strassbourg, France*, 2011.

[113] D. Steffes-lai and T. Clees. Statistical analysis of forming processes as a first step in a process-chain analysis: novel PRO-CHAIN components. *Key Engineering Materials*, 504 – 506:631 – 636, 2012.

[114] D. Steffes-lai, L. Nikitina, and T. Clees. Apparatus and method for editing a process simulation database for a process, Patent, EP2433185, WO2010EP61450, 2010.

[115] D. Steffes-lai, E. Rosseel, and T. Clees. Interpolation methods to compute statistics of a stochastic partial differential equation. *ArXiv e-prints*, 1309.3853, 2013.

[116] D. Steffes-lai, S. Turck, C. Klimmek, and T. Clees. An efficient knowledge based system for the prediction of the technical feasibility of sheet metal forming processes. *Key Engineering Materials*, 554 - 557:2472 – 2478, 2013.

[117] B. Sudret. Global sensitivity analysis using polynomial chaos expansions. *Reliability Engineering and System Safety*, 93:964 – 979, 2008.

[118] D.-Z. Sun, F. Andrieux, and M. Feucht. Damage modelling of trip steel for integrated simulation from deep drawing to crash. In *Proceedings of the 7th European LS-DYNA Users' Conference, Salzburg, Austria*, 2009.

[119] A. E. Tekkaya. State-of-the-art of simulation of sheet metal forming. *Journal of Materials Processing Technology*, 103:14 – 22, 2000.

[120] C.-A. Thole and K. Stüben. Industrial simulation on parallel computers. *Parallel Computing*, 25:2015 – 2037, 1999.

[121] T. Turanyi. Sensitivity analysis of complex kinetic systems. Tools and applications. *Journal of Mathematical Chemistry*, 5:203 – 248, 1990.

[122] H. S. Valberg. *Applied Metal Forming including FEM Analysis*. Cambridge University Press, 2010.

[123] S. Vassilvitskii and D. Arthur. K-means++: The advantage of careful seeding. In *Proceedings of the eighteenth annual ACM-SIAM symposium on Discrete algorithms (SODA),Philadelphia, PA, USA*, pages 1027–1035, 2007.

[124] H. Wendland. *Scattered data approximation*. Cambridge University Press, 2005.

[125] S. Wolff and C. Bucher. Recent developments for random fields and statistics on structures. In *Weimar Optimization and Stochastic Days 9.0*, 2012.

[126] D. Xiu. Fast numerical methods for stochastic computations: a review. *Communications in Computational Physics*, 5(2-4):242–272, 2009.

[127] D. Xiu and J. S. Hesthaven. High-order collocation methods for differential equations with random inputs. *SIAM Journal on Scientific Computing*, 27(3):1118–1139, 2005.

[128] D. Xiu and G. E. Karniadakis. Modeling uncertainty in flow simulations via generalized polynomial chaos. *Journal of Computational Physics*, 187:137–167, 2003.

# List of Figures

203

# List of Tables

# Acronyms

DoE     design of experiments

EPS     effective plastic strain

FD     finite differences
FE     finite element

KL     Karhunen-Loève

MC     Monte Carlo
MSE     mean squared error

NLDR     nonlinear dimension reduction

PCA     principal component analysis
PDE     partial differential equation

RBF     radial basis function
RMSE     root mean squared error

sPDE     stochastic partial differential equation
SVD     singular value decomposition

# List of Symbols

| | |
|---|---|
| $\#S$ | Number of elements of the set $S$ (cardinality) |
| $\mathbf{A}^k$ | Rank-$k$ approximation of matrix $\mathbf{A}$ |
| $|\alpha|_{\max}$ | Maximal absolute eigenvalue of a matrix |
| $b$ | Arbitrary function, $b\colon \mathbb{R}^{\mathrm{Npar}} \to \mathbb{R}^{\mathrm{Ncrit}}$ |
| $\mathcal{B}$ | Boundary operator |
| $C_R$ | Data compression ratio |
| $\mathrm{CL}_i$ | $i$-th cluster |
| $c_{\mathrm{CL}_i}$ | Center of cluster $\mathrm{CL}_i$ |
| $\mathbf{Cov}[W, Z]$ | Covariance of two random variables $W$ and $Z$ |
| $D$ | Measure for design-parameter classification based on full Hessian matrix |
| $D_i$ | Local measure for design-parameter classification based on full Hessian matrix |
| $d_x$ | Damage design-parameters (bi-failure damage model, Equation (7.6)) |
| $\mathrm{Dd}$ | Polynomial degree used in detrending |
| $\delta$ | Constant $> 0$, properly chosen (Equations (6.5), (6.6)) |
| $\delta_n$ | Constant $> 0$, properly chosen (Equation (6.7)) |
| $\mathbf{diag}H$ | Matrix of diagonal parts of Hessian matrices with entries $\mathrm{diag}H_{ij}$ according to mesh node $N_i$ and parameter $P^j$ |
| $\mathrm{diag}H^j$ | Accumulated diagonal part of Hessian matrix w.r.t. design-parameter $P^j$ |
| $\mathbf{diff}_{\mathrm{abs}}$ | Absolute difference between two values |
| $\mathbf{diff}_{\mathrm{abs,i}}$ | Absolute difference between two values in mesh node $N_i$ |
| $\mathbf{diff}_{\mathrm{rel}}$ | Relative difference between two values |

215

| | |
|---|---|
| $\mathbf{diff}_{\text{rel},i}$ | Relative difference between two values in mesh node $N_i$ |
| dim | Dimension of the coordinate space |
| $e_i$ | Element value in mesh element Element$_i$ |
| $\widetilde{e}_i$ | Continuous extension of the element value $e_i$, if Element$_i$ is deleted |
| $\bar{\varepsilon}$ | Effective strain |
| $\bar{\varepsilon}^p$ | Effective plastic strain |
| $\varepsilon_0$ | Pre-strain constant (Swift hardening law) |
| $\varepsilon_\sigma$ | Threshold which denotes the relative amount of variation due to design-parameter variations which is not explained anymore |
| $\varepsilon_{\text{SVD}}$ | Threshold which denotes the amount of relative error due to SVD truncation which is acceptable |
| $\varepsilon_{\text{SVD,abs}}$ | Threshold which denotes the amount of absolute error due to SVD truncation which is acceptable |
| $\mathbf{E}[Z]$ | Expectation (mean value) of a random variable $Z$ |
| Element$_i$ | $i$-th mesh element |
| $\text{err}_1(k)$ | Average approximation error due to truncation of SVD according to Equation (5.11) |
| $\text{err}_2(k)$ | Average approximation error due to truncation of SVD according to Equation (5.12) |
| $F^D$ | Drawing force (process design-parameter) |
| $F^H$ | Hold down force (process design-parameter) |
| $f_Z$ | Probability density function for a random variable $Z$ |
| $F_Z$ | Probability distribution function of a real-valued random variable $Z$ |
| $\mathbf{g}$ | Function vector describing a single processing step on the entire mesh |
| $g_i$ | Function describing a single processing step in mesh node $N_i$ |
| $\widetilde{\mathbf{g}}$ | Approximated function vector describing a single processing step on the entire mesh |
| $\widetilde{g}_i$ | Approximated function describing a single processing step in mesh node $N_i$ |
| $\mathbf{H}$ | Hessian matrix with entries $H_{ij}$ |
| $\mathbf{J}$ | Jacobian matrix with entries $J_{ij}$ |
| $J^j$ | Accumulated Jacobian matrix w.r.t. design-parameter $P^j$ |

| | |
|---|---|
| $\widetilde{\mathbf{J}}$ | Vector of all accumulated linear sensitivities scaled by the corresponding parameter variation range |
| $K$ | Strength coefficient (Swift hardening law) |
| $\mathcal{L}$ | Differential operator |
| $\text{LIClass}_i$ | $i$-th linear importance class |
| $m_{ij}$ | Mapping between distinct processing steps $i$ and $j$ |
| $m_j$ | Mapping within a single processing step $j$ |
| $\mathbf{M}$ | Database containing an ensemble of simulation results according to a single criterion $Y^l$ with entries $M_{ij}$ |
| $\mathbf{M}_i$ | $i$-th row of the database $\mathbf{M}$ |
| $\widehat{\mathbf{M}}$ | Normalized database $\mathbf{M}$ by subtracting the mean of the experiments of each row with entries $\widehat{M}_{ij}$ |
| $\widehat{\mathbf{M}}^k$ | Rank-$k$ approximation of the database $\widehat{\mathbf{M}}$ |
| $\widetilde{\mathbf{M}}$ | Database containing an ensemble of simulation results with coordinates and criteria information about several time steps |
| $\widetilde{\mathbf{M}}(i)$ | Database of the $i$-th timestep |
| $\widetilde{\mathbf{M}}^k$ | Rank-$k$ approximation of the database $\widetilde{\mathbf{M}}$ |
| $\widehat{\widetilde{\mathbf{M}}}$ | Normalized database $\widetilde{\mathbf{M}}$ by subtracting the mean of the experiments of each row |
| $\mu$ | Friction coefficient (process design-parameter) |
| $n$ | Strain-hardening index (Swift hardening law) |
| $N_i$ | $i$-th mesh node |
| $n_i$ | Nodal value in mesh node $N_i$ |
| $n_{\text{del},l}(i)$ | Threshold of nodal values which indicates possible failures in state $i$ of the simulation according to sampling point $\mathbf{P}_l$ |
| $n_{\text{del},\mathbf{M}}(i)$ | Threshold of nodal values which indicates possible failures in state $i$ used in the prediction by a metamodel based on the database $\mathbf{M}$ |
| Nclust | Number of clusters |
| Ncrit | Number of criteria, dimension of criteria space |
| Ndata | Number of data entries (rows of a matrix) |
| Nexp | Number of sampling points (number of simulations) |

| | |
|---|---|
| $\text{NLClass}_i$ | $i$-th nonlinearity class |
| Nnodes | Number of mesh nodes |
| Npar | Number of design-parameters, dimension of design-parameter space |
| Npred | Number of sampling points to evaluate the forecast model (number of predictions) |
| Nq | Number of quantiles |
| Nts | Number of time steps in a simulation |
| $\|\mathbf{X}\|_2$ | Euclidean norm of vector $\mathbf{X}$ |
| $\|\mathbf{X}\|_\infty$ | Maximum norm of vector $\mathbf{X}$ |
| $\|\mathbf{X}\|_p$ | $L_p$ norm of vector $\mathbf{X}$ |
| $\|\mathbf{A}\|_F$ | Frobenius norm of matrix $\mathbf{A}$ |
| $\|\mathbf{A}\|_2$ | Spectral norm of matrix $\mathbf{A}$ |
| $\mathbf{P}$ | Design-parameter vector |
| $P^j$ | $j$-th design-parameter |
| $\mathbf{P}_{\max}$ | Vector of maximal design-parameter values |
| $P^j_{\max}$ | Maximal value of design-parameter $P^j$ |
| $\mathbf{P}_{\min}$ | Vector of minimal design-parameter values |
| $P^j_{\min}$ | Minimal value of design-parameter $P^j$ |
| $\mathbf{P}_l$ | $l$-th sampling point within design-parameter space |
| $\mathbf{P}_{\text{nom}}$ | Nominal design-parameter vector |
| $P^j_{\text{nom}}$ | Nominal value of design-parameter $P^j$ |
| $\mathcal{P}$ | Probability measure |
| $\mathbf{Q}_q(Z)$ | q-quantile for a random variable $Z$ |
| $\widehat{\mathbf{Q}}_q(Z)$ | Estimator of the q-quantile for a random variable $Z$ |
| $\mathbb{R}$ | Space of real numbers |
| $R_i$ | Anisotropy coefficient (material design-parameter), $i$ is the angle between the axis of the specimen and the rolling direction |
| $s_i$ | Surface area assigned to element $\text{Element}_i$ |
| $s_{\text{c}}$ | Compressed size of data |
| $s_{\text{unc}}$ | Uncompressed size of data |
| S | Set of sampling points |
| $\boldsymbol{S1}$ | Local sensitivity matrix with entries $S1_{ij}$ |
| $S1^j$ | Accumulated sensitivity measure of design-parameter $P^j$ |
| $\widetilde{S1}^j$ | Accumulated sensitivity measure of design-parameter $P^j$ scaled to $[0,1]$ |
| $\boldsymbol{S2}$ | Matrix of local nonlinear influences with entries $S2_{ij}$ |

| | |
|---|---|
| $S2^j$ | Accumulated measure of nonlinear influences of design-parameter $P^j$ |
| $\widetilde{S2}^j$ | Accumulated measure of nonlinear influences of design-parameter $P^j$ scaled to $[0, 1]$ |
| $\sigma_{\mathbf{P}}$ | Vector of standard deviations from corresponding nominal design-parameter vector $\mathbf{P}_{\mathrm{nom}}$ |
| $\sigma_{P^j}$ | Standard deviation from nominal design-parameter value of design-parameter $P^j$ |
| $t$ | Metal sheet thickness |
| $\mathrm{TI}(k)$ | Proportion of overall variation explained by parameters in the first $k$ total importance classes |
| $\mathrm{TI}_i(k)$ | Proportion of overall variation explained locally in mesh node $N_i$ by parameters in the first $k$ total importance classes |
| $\mathrm{TI}_{ij}$ | Local total sensitivity of $P^j$ in mesh node $N_i$ |
| $\widetilde{\mathrm{TI}}^j$ | Accumulated total sensitivity of design-parameter $P^j$ scaled to $[0, 1]$ |
| $\mathrm{TIClass}_i$ | $i$-th total importance class |
| trace | Trace of a matrix |
| ts | State, current timestep |
| $T_{\mathrm{trans}}$ | Triaxiality (bi-failure damage model) |
| $\mathbf{Var}[Z]$ | Variance of a random variable $Z$ |
| $w_k$ | Weight of radial basis function metamodel |
| $\mathbf{X}$ | General vector of coordinates |
| $\mathbf{X}^I$ | Vector of input coordinates, $\mathbf{X}^I \in \mathbb{R}^{\dim}$, usually dim $= 3$ |
| $\mathbf{Y}$ | Criteria vector |
| $Y^j$ | $j$-th criterion |
| $\widetilde{Y}^j$ | Approximated $j$-th criterion |
| $\mathbf{Y}_l$ | $l$-th criteria vector corresponding to sampling point $\mathbf{P}_l$ |
| $Z$ | Random variable |